SpringerBriefs in Applied Sciences and Technology

SpringerBriefs present concise summaries of cutting-edge research and practical applications across a wide spectrum of fields. Featuring compact volumes of 50 to 125 pages, the series covers a range of content from professional to academic.

Typical publications can be:

- A timely report of state-of-the art methods
- An introduction to or a manual for the application of mathematical or computer techniques
- A bridge between new research results, as published in journal articles
- A snapshot of a hot or emerging topic
- An in-depth case study
- A presentation of core concepts that students must understand in order to make independent contributions

SpringerBriefs are characterized by fast, global electronic dissemination, standard publishing contracts, standardized manuscript preparation and formatting guidelines, and expedited production schedules.

On the one hand, **SpringerBriefs in Applied Sciences and Technology** are devoted to the publication of fundamentals and applications within the different classical engineering disciplines as well as in interdisciplinary fields that recently emerged between these areas. On the other hand, as the boundary separating fundamental research and applied technology is more and more dissolving, this series is particularly open to trans-disciplinary topics between fundamental science and engineering.

Indexed by EI-Compendex, SCOPUS and Springerlink.

Maya Genisa · Solehuddin Shuib ·
Zainul Ahmad Rajion

Biomechanics of Dental Implants

3-Dimensional Bone Assessment Using CBCT from Laboratory to Clinics

Maya Genisa
Post Graduate School
YARSI University
Jakarta, Indonesia

Solehuddin Shuib
School of Mechanical Engineering
College of Engineering
Universiti Teknologi MARA
Shah Alam, Selangor, Malaysia

Zainul Ahmad Rajion
Department of Oral Maxillofacial Surgery
and Oral Diagnosis, Kulliyyah of Dentistry
International Islamic University Malaysia
(IIUM)
Kuantan, Malaysia

ISSN 2191-530X   ISSN 2191-5318 (electronic)
SpringerBriefs in Applied Sciences and Technology
ISBN 978-981-96-6022-3   ISBN 978-981-96-6023-0 (eBook)
https://doi.org/10.1007/978-981-96-6023-0

© The Editor(s) (if applicable) and The Author(s), under exclusive license to Springer Nature Singapore Pte Ltd. 2025

This work is subject to copyright. All rights are solely and exclusively licensed by the Publisher, whether the whole or part of the material is concerned, specifically the rights of translation, reprinting, reuse of illustrations, recitation, broadcasting, reproduction on microfilms or in any other physical way, and transmission or information storage and retrieval, electronic adaptation, computer software, or by similar or dissimilar methodology now known or hereafter developed.
The use of general descriptive names, registered names, trademarks, service marks, etc. in this publication does not imply, even in the absence of a specific statement, that such names are exempt from the relevant protective laws and regulations and therefore free for general use.
The publisher, the authors and the editors are safe to assume that the advice and information in this book are believed to be true and accurate at the date of publication. Neither the publisher nor the authors or the editors give a warranty, expressed or implied, with respect to the material contained herein or for any errors or omissions that may have been made. The publisher remains neutral with regard to jurisdictional claims in published maps and institutional affiliations.

This Springer imprint is published by the registered company Springer Nature Singapore Pte Ltd.
The registered company address is: 152 Beach Road, #21-01/04 Gateway East, Singapore 189721, Singapore

If disposing of this product, please recycle the paper.

Preface

Introduction

The application of CBCT for 3-D bone assessment in dental implant planning is an essential and fast growth in dentistry. Various methods for quantifying bone density, cortical thickness, and other relevant parameters have been developed and tested. On the other side, the use of CBCT-based biomechanical simulations to predict implant stability and potential complications is still a challenging task and still has space for development.

The book is structured to guide both clinicians and researchers through the intricate relationship between biomechanics and dental implant success. We begin by exploring the fundamental principles of biomechanics, including stress, strain, and bone adaptation. This knowledge is essential for understanding the forces acting on dental implants and the subsequent response of the surrounding bone tissue.

The intricacies of CBCT technology, from image acquisition to analysis techniques, and the advantages and limitations of CBCT in comparison to other imaging modalities are discussed to provide practical guidelines for selecting appropriate imaging protocols.

The final chapters of the book bridge the gap between laboratory research and clinical practice. We present case studies demonstrating the clinical application of CBCT-guided implant planning and discuss the impact of 3-D bone assessment on treatment outcomes. Furthermore, we explore emerging trends in CBCT technology and future directions in the field of dental implant biomechanics.

Purpose

This book aims to provide a comprehensive overview of the biomechanical principles underlying dental implant treatment and the pivotal role of Cone Beam Computed Tomography (CBCT) in 3-D bone assessment. As dental implant therapy continues to evolve, the demand for precise and accurate bone evaluation has become paramount. CBCT technology has emerged as a valuable tool in this regard, offering detailed, non-invasive imaging of the jawbone.

Acknowledgments

This book would not have been possible without the invaluable contributions of many individuals. I would like to express my sincere gratitude to my supervisor during my Ph.D. candidacy, Prof. Dr. Zainul Ahmad Rajion and Prof. Ir. Ts. Dr. Solehuddin Shuib, for their guidance, expertise, and unwavering support that have been instrumental in my research and development, and the dental implant community patients: Their participation and dedication to advancing the field and their willingness to share knowledge have been invaluable.

I would also like to acknowledge the top management of YARSI University, Prof. Dr. H. Fasli Jalal Ph.D. and Prof. Dr. Jurnalis Uddin, P.A.K., for their continuous support and encouragement.

Finally, we would like to dedicate this book to our family, whose love and encouragement have been a constant source of inspiration.

Outline

The book is organized into nine chapters. Chapter 1 introduces the biomechanics of dental implants and explores 3-D bone assessment using CBCT from laboratory to clinical applications. Chapter 2 delves deeper into the biomechanics of dental implants, specifically focusing on three-dimensional bone assessment. Chapter 3 addresses digital imaging, implant stability, and Finite Element Analysis (FEA) of study preparations. Chapter 4 discusses biomechanical assessments based on clinical measurements. Chapter 5 examines the application of CBCT data for monitoring bone quality and implant stability, as well as their correlation. Chapter 6 focuses on the biomechanical assessment of dental implants using FEA. Chapter 7 explores biomechanical assessments of patients with high implant stability, while Chap. 8 discusses those with low implant stability. Finally, the last chapter (Chap. 9) covers the fundamentals of 3-D bone assessment using CBCT, bridging laboratory research and clinical practice.

Conclusion

Finally, we expect that this book is intended to serve as a valuable resource for dental practitioners, researchers, and students interested in the field of dental implantology.

By providing a solid foundation in biomechanics and CBCT technology, we hope to empower readers to make informed decisions and optimize the outcomes of dental implant treatment.

Jakarta, Indonesia Dr. Maya Genisa
Shah Alam, Malaysia Prof. Ir. Ts. Dr. Solehuddin Shuib
Kuala Lumpur, Malaysia Prof. Dr. Zainul Ahmad Rajion

Contents

1 **Biomechanics of Dental Implant: Three-Dimensional Bone Assessment Using CBCT from Laboratory to Clinic** 1
  1.1 Background 1
  1.2 Statement of Problems 2
  1.3 Significant of Study 4
  1.4 Objective of the Study 4
  1.5 Structure of the Book 4
  References 5

2 **Biomechanics of Dental Implant: Three-Dimensional Bone Assessment** 7
  2.1 Introduction 7
  2.2 Bone Formation 7
  2.3 Bone Quality and Quantity Classification 8
  2.4 Procedures of Implant Placement 10
  2.5 Methods of Implant Stability Measurement in Dentistry 11
  2.6 Tensional Test Method 11
    2.6.1 Removal Torque Method 12
    2.6.2 Cutting Resistance or Insertion Torque Method 12
    2.6.3 Resonance Frequency Analysis (RFA) Method 13
  2.7 Dental Imaging Technology 14
    2.7.1 Computerized Tomography (CT) Scan 14
    2.7.2 Cone Beam Computer Tomography (CBCT) 15
  2.8 Biomechanical of Dental Implant System 16
  2.9 Application of Finite Element Analysis for Dental Implant Assessment 18
  References 20

3 Digital Imaging and Implant Stability, and Finite Element Analysis of Study Preparation ... 25
- 3.1 Introduction ... 25
- 3.2 Ethical Approval ... 25
  - 3.2.1 Study Design ... 25
  - 3.2.2 Reference Population ... 27
  - 3.2.3 Sample Size ... 27
  - 3.2.4 Statistical Analysis ... 27
- 3.3 Procedure of Implant Placement ... 28
- 3.4 Procedure of Dental Impression ... 29
- 3.5 Procedure of Crown Installation ... 31
- 3.6 Dental Imaging and Implant Stability Measurement ... 33
  - 3.6.1 Validation of CBCT Scanning on Phantom ... 33
  - 3.6.2 Validation of CBCT Scanning Using CT Scanning ... 35
  - 3.6.3 Measurement of Bone Density Using MIMICS Software ... 36
  - 3.6.4 Measurement Implant Stability Using Resonance Frequency Analysis ... 36
- 3.7 Finite Element Analysis (FEA) Study Preparation ... 37
  - 3.7.1 FEA Workflow ... 38
  - 3.7.2 Material Assignment and Boundary Conditions ... 39
  - 3.7.3 FEA Simulations ... 40
- References ... 41

4 Biomechanical Assessment Based on Clinical Measurement ... 43
- 4.1 Accuracy and Repeatability Assessment ... 43
- 4.2 Repeatability Measurement of Bone Density Using MIMICS Software ... 44
- 4.3 Validation Density Measurement of CBCT Using Phantom ... 46
- 4.4 Justification of Density Measurement Using CT Scanning ... 47
- 4.5 Discussion ... 49
- References ... 54

5 Application of CBCT Data for Bone Quality and Implant Stability Monitoring and Correlation ... 55
- 5.1 Application of CBCT Data for Bone Quality and Quantity Assessment ... 55
- 5.2 Bone Density Evaluation of Pre- and Post-Crown ... 55
- 5.3 Cortical Thickness and Available Space Measurement ... 57
- 5.4 Discussions ... 59
- 5.5 Implant Stability Monitoring and Correlation ... 60
  - 5.5.1 Implant Stability Measurement ... 60
  - 5.5.2 Correlation Between Bone Quality/Quantity and Implant Stability ... 61
  - 5.5.3 Discussions ... 63
- References ... 65

6	**Biomechanical Assessment of Dental Implant Using Finite Element Analysis (FEA)**		67
	6.1 Introduction		67
	6.2 Mechanism of Stress Distribution on Simple Model		68
		6.2.1 Vertical Force Simulation: Pre- and Post-Crown Condition	68
		6.2.2 Horizontal Force Simulation: Pre- and Post-Crown	71
		6.2.3 Removal Torque Simulation	73
		6.2.4 Summary of Result	73
	6.3 Effect of Cortical Thickness and Friction Coefficient on Stress Distribution		75
		6.3.1 Effect of Cortical Thickness on Stress Distribution	75
		6.3.2 Effect of Friction Coefficient on Stress Distribution and Micro Motion	76
		6.3.3 Summary of Result	78
	References		79
7	**Biomechanical Assessment of Patient with High Implant Stability**		81
	7.1 Biomechanical Assessment of Patients with High Implant Stability		81
		7.1.1 Behaviour of Stress Distribution on High Implant Stability Patient	81
		7.1.2 Micro Motion of Implant and Neighbour Teeth	84
	7.2 Summary of Result		85
	7.3 Biomechanical Assessment of Patient with Moderate Implant Stability		85
		7.3.1 Behaviour of Stress Distribution on Moderate Implant Stability Patient	86
		7.3.2 Micro Motion on Moderate Implant Stability Patient	87
	7.4 Summary of Result		87
8	**Biomechanical Assessment on Patient with Low Implant Stability**		91
	8.1 Biomechanical Assessment on Patient with Low Implant Stability		91
		8.1.1 Behaviour of Stress Distribution on Low Implant Stability Patient	92
		8.1.2 Micromotion on Low Implant Stability Patient	94
		8.1.3 Summary of Result	95
	8.2 General Discussion on Biomechanical Evaluation of Dental Implant		96
	Reference		98

9 Fundamental of Three-Dimensional Bone Assessment Using CBCT from Laboratory to Clinics 99
    9.1   Conclusion 99
    9.2   Future Work 101

Index 103

List of Figures

Fig. 2.1	Jawbone structure obtained from CBCT image	8
Fig. 2.2	Anatomy of bone: cortical and trabecular pictures (modified: Dent-Wiki.co http://www.dent-wiki.com/dental_technology/alveolar-bone-structure/)	17
Fig. 3.1	Workflow of research consists of in vivo and in vitro studies	26
Fig. 3.2	**a** Loss of the molar tooth, **b** placement implant insertions, and **c** torquing wrench/screwdriver instrument is being used to tighten the implant	28
Fig. 3.3	**a** Healing screw, **b** condition after insertion of the healing screw, and **c** condition after suture	28
Fig. 3.4	**a** Suture after implant placement, and **b** after 3 months, replacement healing screw and placement of the gingival former screw	29
Fig. 3.5	**a** Condition of implant 1 week after placement gingival former screw, unscrew the gingival screw for placement the transfers coping into the implant, and **b** transfers coping with screw partly intruded	30
Fig. 3.6	**a** Dental impression will be used later when the dentist is ready to make the crown, **b** impression of the lower teeth with the transfers coping in place is now sent to the lab, and **c** screwing the healing cap back on to the implant for two weeks	30
Fig. 3.7	Crown will be ready in 2 weeks. Before that crown installed into patient, the finishing, refinement, and repositioning with maxillaries and mandibular teeth are performed	31
Fig. 3.8	**a** Abutment and **b** crown ready for installation	31
Fig. 3.9	**a** Unscrewing the healing cap from the implant and **b** installed abutment into the implant	32
Fig. 3.10	**a** After the crown is installed into the abutment and **b** after cementation of the crown as the final stage of implant treatment	32

Fig. 3.11	**a** Model 711-HN and **b** the position on the CBCT scanning	34
Fig. 3.12	CBCT scanning result of phantom 711-HN model and its density measurement on MIMICS software at 8 mm level (apically) from CEJ	35
Fig. 3.13	CT scanning of phantom with different angles: **a** 0°, **b** 15°, and **c** 30°	36
Fig. 3.14	**a** Implant with a SmartPeg to measure stability implant using RFA and **b** measurement of RFA from the buccal and lingual side	37
Fig. 3.15	Workflow of FEA study	38
Fig. 3.16	Process of segmentation from CBCT data to construct 3-D objects	39
Fig. 3.17	Simulation of different loading: **a** vertical loading, **a** horizontal loading, and **c**. Removal torque, at pre-crown and post-crown conditions. Arrows show a force/torque location	41
Fig. 4.1	Measured density in HU from CBCT data using **a** 2-D method and **b** 3-D method	45
Fig. 4.2	Density measurement of phantom on CBCT data at: **a** cortical bone and **b** enamel	47
Fig. 4.3	Relation between true density and greyscale of CBCT of the phantom **a** linear regression, **b** logarithmic regression	48
Fig. 4.4	Curve estimation of density—HU relation from CT scanning: **a** linear and **b** logarithmic	50
Fig. 4.5	Cross-plot between greyscale of CBCT and HU of CT	51
Fig. 4.6	Difference between CT and CBCT: **a** configuration of fan beam (left) and cone beam source (right) and **b** effect of misalignment on CT (left) and CBCT (right)	52
Fig. 5.1	**a** Standing position of the patient during CBCT scanning and **b** density estimation from CBCT data in Mimics software	56
Fig. 5.2	Density of site implant during monitoring stage. **a** In greyscale and **b** In gr/cc	57
Fig. 5.3	Measured cortical thickness from CBCT data	58
Fig. 5.4	Width and height of jaw measured from CBCT data	58
Fig. 5.5	Volume of jaw around implant site, measured from CBCT data	59
Fig. 5.6	**a** Implant insertion surgery, **b** implant stability measurement using RFA Osstell mentor device	61
Fig. 5.7	Plot between density and primary implant stability that are measured in stage 1	62
Fig. 5.8	Primary implant stability cross-plot against **a** bone width, **b** bone height, **c** cortical thickness, and **d** angle insertion	63
Fig. 5.9	Progress of implant stability and density changes during the monitoring period (**a**). Stage 2, and (**b**) Stage 3	64

List of Figures xv

Fig. 6.1	Geometry of each component of a simple model. **a** Complete model, **b–c** geometry of cortical bone, **d** geometry of implant-crown, **e** size of implant, and **f** size of implant and crown derived from CBCT data	69
Fig. 6.2	Meshing of dental implant system **a** 3-D view, and **b** cross-section view. This simulation was solved with automatic meshing: 364,998 nodes, 255,465 mesh, and a minimum edge length was 0.341 mm	70
Fig. 6.3	Fixed support for Finite Element Analysis is at the bottom of the model	70
Fig. 6.4	**a** Push-out simulation with a 200 N vertical force is loaded into the implant dental system, pre-crown (left) and post-crown conditions (right), and **b** Von Mises stress distribution in 3-D view resulted from FEA simulation	71
Fig. 6.5	Horizontal force simulation of pre- and post-crown, a). location of horizontal force (200 Newton), and b). Von Mises stress distribution in 3-D view of pre- and post-crown	72
Fig. 6.6	Removal torque simulation for pre- and post-crown: **a** Location of removal torque, and **b** stress distribution in 3-D view	74
Fig. 6.7	Models of dental implant systems with different cortical thicknesses of 2.30, 2.85, 3.53, and 3.93 mm	76
Fig. 6.8	Von Mises stress in a 3-D view of each model with different cortical thicknesses	77
Fig. 6.9	Von Mises stress at different probes for different friction coefficients	77
Fig. 7.1	Model of dental implant system derived from CBCT of the patient with high implant stability. **a** Components of the dental implant system consist of cortical and trabecular of the jawbone, two neighbour teeth, implant body, and crown, **b** pre-crown model, and **c** post-crown model	82
Fig. 7.2	FEA simulations for different types of loading on pre- and post-crown conditions. **a** Vertical force, **b** horizontal force, and **c** removal torque simulation	83
Fig. 7.3	Geometry of pre- and post-crown of patients with moderate implant stability, **a** components of the dental implant system, and **b** complete meshing of the dental implant system for pre- and post-crown conditions (automatic meshing: 24,384 nodes, 91,819 elements)	86
Fig. 7.4	**a** Von Mises stresses on moderate implant stability model generated from different types of loading. **a** Vertical force, **b** horizontal force, and **c** removal torque simulation at pre- and post-crown conditions	88

Fig. 8.1	Model of dental implant system of a patient with low implant stability. **a** Components of the dental implant system, and **b** pre- and post-crown model (automatic meshing: 5655 nodes, 17,856 elements)	92
Fig. 8.2	Stress distribution on the model of low implant stability patient. **a** Vertical loading, **b** horizontal loading, and **c** removal torque at pre- and post-crown conditions	93

List of Tables

Table 2.1	Advantages and disadvantages of CBCT (Gupta 2016; Pascual and Morale 2015)	16
Table 3.1	Evaluation measurement schedule during implant placement	33
Table 3.2	Density and material of elements of CIRS Model 711-HN	34
Table 3.3	Material properties for material assignment during FEA study	39
Table 4.1	Descriptive statistics: mean, standard deviation, and minimum and maximum of bone density values, defined as grey density values (VV)	44
Table 4.2	Significance difference of each group	44
Table 4.3	Repeatability and significant difference between measurement 1 and measurement 2 in 2-D and 3-D methods	46
Table 4.4	Measured density based on CT data and true density of the object	49
Table 5.1	Descriptive statistics of available space around the site implant	59
Table 7.1	Micro motion of implant system due to a vertical force, horizontal force, and removal torque of high implant stability patient	84
Table 7.2	Micromotion of the implant system of the patient with moderate implant stability	89
Table 8.1	Micromotion of an implant system of the patient with low implant stability	95

Chapter 1
Biomechanics of Dental Implant: Three-Dimensional Bone Assessment Using CBCT from Laboratory to Clinic

1.1 Background

A dental implant is defined as a device of biocompatible material(s) placed into the mandibular or maxillary bone to replace the edentulous tooth. In addition, it is used also to improve appearance, and masticatory function and prevent changes in dental arch dimension. Thus, dental implants not only improve the convenience of the patient, it is also able to protect the remaining natural teeth, no bone loss and restore facial skeletal structure (Staden et al. 2006).

Statistically, it was shown that the use of dental implants to restore missing teeth has become increasingly widespread over the past two decades (Turkyilmaz and Mcglumphy 2008). The statistics also showed that the success rate of dental implants is over 95% when the implants are designed, manufactured, and placed correctly. Staden et al. (2006) calculated the survival rate at 15 years about 90% which becomes an advantage in implant treatment as low-risk treatment.

While the success rate of dental implant treatment is high, the compatibility of the installed implant into the jawbone system might generate a problem because the forces conveyed by implant devices differ from those conveyed by natural teeth; thus, they require adaptation from the jawbones. Maximum adaptation will be determined by the success of how the implant is integrated into the bone as the so-called osseointegration process. Biomechanical becomes a most important issue in implant dentistry.

The research on implant dentistry grew rapidly after Lekholm and Zarb (1985) published his findings as a theory of osseointegration. In his theory, osseointegration is defined as a direct structural and functional connection between living bone and the surface of a loading-bearing implant. Many aspects have been investigated, especially the aspect related to measuring the stability of implant after implant insertion and correlating it with the mechanism of the internal process of bone to build optimum osseointegration. It still becomes a big challenge to solve.

Meredith (1998) mentioned that some factors affect the success of the implant such as the primary implant stability which comes from mechanical engagement with cortical bone during implant placement, osseointegration, implant placement technique, and local bone quality and quantity. Another important determinant for a successful implant is secondary stability. Secondary stability offers biological stability through bone regeneration and remodelling (Atsumi et al. 2007). Continuous monitoring in an objective manner of the status of implant stability will be important to be established to ensure implant success in future.

Lack of osseointegration during implant dental rehabilitation has been reported. It is because of some reasons that might happen during treatment such as infection processes or inadequate load protocols. Incorrect placement technique and the shape of the implant surface will reduce the coupling between bone and implant surface which can produce spaces where bacteria could grow. On the other hand, the mechanism of daily mastication also may produce large loading that promotes the mobility of the implant and holes. However, if the generated loading is too small, it is not sufficient to stimulate osteoblast activity, and hence the osseointegration might be delayed. The mechanism of loading and its effect on progressing the rehabilitation become a most important task and the behaviour of the jaw system during treatment, especially the stress distribution due to loading still unclear and still needs further investigation.

More questions about stress distribution are still not solved, how bone reacts to the generated stress during loading and its relation between generated stress with a mechanical and hormonal response and remodelling/osseointegration, which is still unclear. Therefore, it is important to study the stress patterns distribution and correlate it with the osseointegration process. However, because of the limitation on available clinical instruments that can measure the stress distribution directly, alternative analysis based on numerical computation would be the ultimate method to understand the biomechanical mechanism of the implant dental system.

Finite element analysis (FEA) is a numerical method, which could solve complex mechanical problems into elements, and has been well accepted for investigating the behaviour of stress in dentistry. Various loadings can be examined in different models of in vitro or in vivo situations. This study, a prospective observational study is conducted to integrate the clinical measurements of implant stability using Resonance Frequency Analysis (RFA), and site implant measurement using CBCT with numerical studies through FEA to assess the biomechanical of dental implants comprehensively in 3-D.

1.2 Statement of Problems

Perfect osseointegration between implant and bone that is indicated by high implant stability is a main goal to be achieved during dental implant treatment. Internal and external factors affect the process of achieving early osseointegration. The external factors such as the size of the implant, technical, and protocol of implant placement,

1.2 Statement of Problems

coupling between implant and bone, stress experienced during treatment, and internal factors such as the quality and quantity of bone, health of mouth environment and internal activity of bone as a response to the external loading, are important factors that determine the success of dental implant treatment. However, the exact relation between those factors and with activity of bone during the healing process is still unclear. There is a need to do comprehensive regular monitoring to measure implant stability which has not been investigated.

CBCT imaging is a new technology in dentistry, especially in Hospital USM. The accuracy of CBCT in determining the geometry of the jawbone with a high-resolution image and the ability to define the geometry of the object up to millimetre scale. However, the accuracy of this method in estimating density is still questionable. More justifications for determining the density using this method still need to be validated and calibrated to gain more confidence in interpreting the data.

Clinically, the internal activity of bone or the ability of osteoblast to respond to the impact of surgery during implant placement is determined by the intensity of stress received during daily loading from the mastication process. However, there is no clinical instrument that can be used to measure directly the effect of those loading including the generated stress around the implant site which can be correlated with an internal bone activity (osseointegration process). It is important to have a method for evaluating the stress distribution due to various external loads in different conditions.

In advance, the developed method should be able to support the clinical measurement, and hence the result of this method can be used as an early warning to maintain the progress of sustainable dental implants. A combination of numerical analysis and clinical measurement needs to be conducted to broaden the use of CBCT data for a better understanding of the biomechanical evaluation of the jaw system, and it would be a bridge between laboratory and clinical assessment.

Using 2-D images as tools for evaluating the jawbone can provide only the image for identifying the static parameters such as density and availability of space for the implant site evaluation. However, when the dynamic properties such as the evaluation of stress distribution which usually occurs three-dimensionally, this method is not supported. On the other side, CBCT scanning can provide a 3-D image of the jaw. Then the 3-D model of the jaw can be generated for further study on biomechanical evaluation and simulation by using either in vivo or in vitro data.

The healing process of implant treatment includes the pre- and post-crown condition. However, the comparative study of biomechanical evaluation of implant systems in pre- and post-crown conditions has not been established yet. Hence, the comparison of dynamic behaviour such as stress distribution for both conditions is still not evaluated. This study gave a better understanding of the mechanism of stress distribution and was able to give feedback to dentists from an engineering perspective.

1.3 Significant of Study

This study could provide reliable measurements of bone density based on CBCT during the monitoring period, while also introducing a new workflow for biomechanical evaluation that integrates clinical measurements and numerical studies. It holds particular significance for the craniofacial research community at USM by supporting imaging technology and workflows for biomechanical assessments based on CBCT data, which can be extended to other imaging technologies such as CT. Additionally, the evaluation of mandibular density using 3-D CBCT offers clinicians a more accurate prediction of bone density variations, aiding in the pre-evaluation of implant site availability.

1.4 Objective of the Study

The specific objectives of this study are:

- To measure the accuracy and repeatability of CBCT scanning in evaluating the bone quality and quantity of dental implant patients.
- To analyse the relationship between bone quality and quantity measured from CBCT with implant stability and its classification that are measured by using RFA.
- To determine bone density changes in the jawbone during dental implant treatment.
- To analyse the effect of cortical thickness and friction coefficient on stress distributions and micromotion of dental implant system during loading in pre- and post-crown conditions.
- To measure the stress distribution and micro motion due to various loadings on pre- and post-crown conditions of in vivo data.

1.5 Structure of the Book

The book is organized into nine chapters. The first chapter introduces the biomechanics of dental implants and explores 3-D bone assessment using CBCT from laboratory to clinical applications. Chapter two delves deeper into the biomechanics of dental implants, specifically focusing on three-dimensional bone assessment. The third chapter addresses digital imaging, implant stability, and finite element analysis (FEA) of study preparations. Chapter 4 discusses biomechanical assessments based on clinical measurements. Chapter 5 examines the application of CBCT data for monitoring bone quality and implant stability, as well as their correlation. Chapter 6 focuses on the biomechanical assessment of dental implants using FEA. Chapter 7 explores biomechanical assessments of patients with high implant stability, while

Chap. 8 discusses those with low implant stability. Finally, the last chapter covers the fundamentals of 3-D bone assessment using CBCT, bridging laboratory research and clinical practice.

References

M. Atsumi, S.H. Park, H.L. Wang, Methods used to assess implant stability: current status. Int. J. Oral & Maxillof. Implant. **22**(5), 743–754 (2007)

U. Lekholm, G.A. Zarb, Patient selection and preparation, in *Tissue Integrated Prostheses: Osseointegration in Clinical Dentistry* (Quintessence Publishing, 1985)

N. Meredith, Assessment of implant stability as a prognostic determinant. Int. J. Prosthodont. **11**(5), 491–502 (1998)

V. Staden, H. Guan, Y.C. Loo, Application of finite element method in dental implant research. Comput. Methods Biomech. Biomed. Eng. **9**(4), 257–270 (2006)

I. Turkyilmaz, E.A. McGlumphy, Influence of bone density on implant stability parameters and implant success: a retrospective clinical study. BMC Oral Health **8**, 32 (2008)

Chapter 2
Biomechanics of Dental Implant: Three-Dimensional Bone Assessment

2.1 Introduction

The main goal of this chapter is to provide the background theory of biomechanical dental implant systems the related research and technology of dental implants and numerical study through Finite Element Analysis (FEA). The explanation includes a review of the technology related to implant stability which includes a brief description of the molar of the mandibular and maxilla, the fundamental mechanism of FEA and Resonance Frequency Analysis (RFA) and the density estimation and its effect on implant stability. In the end, this chapter also discussed the technology of dental imaging including the explanation of Cone Beam Computed Tomography (CBCT) instrument and its application in dental imaging.

2.2 Bone Formation

Bone is the main part of the body that supports the body as a framework; it is essential for the success of dental implant treatment. The formation of bone consists of main parts which are cells and bone matrix. The bone matrix itself consists of inorganic components that cover about 69% hydroxyapatite and 22% organic components that consist of collagen as a major constituent (Kini and Nandeesh 2012). In a physiological view, bone is composed of supporting cells that are osteoblasts and osteocytes and remodelling cells that are osteoclasts. These components are responsible for the dynamic process that consists of the modelling and remodelling process in the bone.

Bone is characterized by its rigidity, hardness, and the degree of the dynamic process in the bone itself such as the modelling and remodelling process. Based on these characteristics, bone can be categorized into two components that are cortical bone and trabecular bone. The cortical bone is dense, and solid surrounds the marrow

Fig. 2.1 Jawbone structure obtained from CBCT image

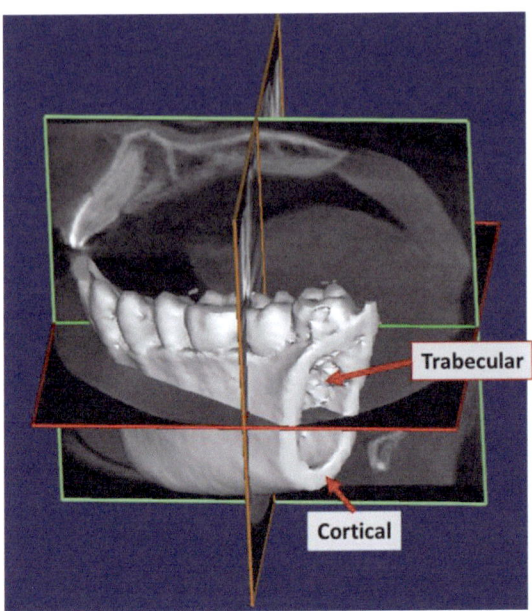

space while the trabecular bone is the honeycomb-like network that consists of trabecular plates and rods interspersed (Kini and Nandeesh 2012). The slicing of cortical bone and trabecular bone is shown in Fig. 2.1.

In the mandible, cortical bone is identified as a clear white in the CBCT image without a trabecular pattern. Meanwhile, trabecular bone is defined as a bone in between two cortical plates. The proportion between trabecular and cortical bone defines the quality of bone in the implant site. If the proportion of cortical bone is higher than trabecular, the bone quality of the implant site is better.

2.3 Bone Quality and Quantity Classification

Bone behaviour can be characterized based on bone quality and quantity. It is a vital factor in achieving osseointegration (Turkyilmaz and McGlumphy 2008). Turkyilmaz and Mcglumphy (2008) found that the survival rate of implants in the mandible is higher than maxilla. This is because the quality of bone in the mandible is higher compared with the maxilla, especially for the posterior maxilla. Bone quality is defined as the ability of bone to withstand a wide range of loading without breaking (Sievänen et al. 2007). In other terms, Lester (2005) defined bone quality as a sum of the total of the bone characteristics that influence the bone's resistance to fracture. Bone quality and quantity affect the survival rate of implants.

2.3 Bone Quality and Quantity Classification

Lekholm and Zarb (1985) have started quality and quantity assessment of bone. In their method, bone quality is scaled between 1 and 4, which is based on radiographic assessment, and the sensation of resistance experienced by the surgeon when preparing the implant site. Based on panoramic radiographs, they classified bone quality as: Type I which has homogenous compact bone, Type II has a thick layer of compact bone surrounding trabecular bone, Type III has a thin layer of cortical surrounding dense trabecular bone, and Type IV has a thin layer of cortical surrounding low-density trabecular bone.

Classification of bone density was introduced by Misch and Abbas (2008). In their classification, bone density was classified based on subjective perception of drilling resistance into four types: D1, D2, D3, and D4. D1 type has a primarily dense cortical bone, the D2 type is dense to the porous cortical bone, the D3 type has a thin cortical and fine trabecular, and D4 has a fine trabecular.

Other researchers, Linkow et al. (2012) classified bone based on structure into three groups: Class I, Class II, and Class III. Class I is identified as bone with spaced trabecular with a small cancellous space. Class II has a larger cancellous, and the uniformity of osseous pattern is less. Class III is identified as bone with a structure of large marrow-filled space between the trabecular.

Most of the explained techniques in determining the bone quality and quantity are based on subjective experience which has difficulties quantifying in general. This classification will depend on the person-to-person who does the evaluation. It is interesting if there is an alternative method that can be used to evaluate bone quality and quantity quantitatively. Hence, the evaluation of bone can be done without personal interference. Because the bone system is a 3-D object, it is interesting if the 3-D image technology can be optimized as a tool for bone quality and quantity assessment. However, the linearity of available 3-D image technology is another issue. Hence, the evaluation of 3-D image parameters as density indicators needs to be tested.

The classifications of bone based on subjective parameters have recently been questioned due to poor objectivity and reproducibility (Shapurian et al. 2006; Al-Nakib 2014). Other techniques such as cutting resistance are performed during surgery; it can be correlated with bone density as assessed by microradiography (Sençimen et al. 2011; Friberg et al. 1995). Direct measurement of insertion torque also showed that there is a strong correlation between insertion torque with density (Bayarchimeg et al. 2013; Khayat et al. 2013).

Micro-computed tomography (Micro CT) is a new method in dental imaging which has very high resolution. This method also can be used to assess jawbone in three-dimensionally. The assessment includes bone volume, density, trabecular thickness, trabecular separation, trabecular number, and structural model index (Fuller et al. 2015). The 3-D morphometric data acquired through micro CT also can be correlated with conventional bone assessment methods (Fanuscu and Chang 2004; Norton et al. 2001; Faot et al. 2015).

Cone Beam Computed Tomography (CBCT) with low dosages, fast scanning time, non-destructive, and the ability to produce high resolution is an alternative dental imaging technology in dentistry. Aranyarachkul et al. (2005) demonstrated that

CBCT could be an alternative diagnostic method for density evaluation, especially since the radiation dosage of CBCT is much lower than computed tomography (CT). Not only it can be correlated with bone density but also it can be correlated with cutting resistance (CR) values as obtained at the time of implant placement.

Chopra (2010) correlated cone beam computerized tomography (CBCT) bone density values with cutting resistance (CR) as obtained during implant placement. Also, the optimization of CBCT for density assessment and correlation with the direction of tooth movement has been introduced by Chang et al. (2012). Other researchers used CBCT as a density prediction tool and correlated it with other measured parameters such as primary implant stability (Marquezan et al. 2012; Isoda et al. 2012).

CBCT offers a high spatial resolution, fast scanning time, and low dosage of exposure is an alternative technique in bone quality and quantity assessment technique based on imaging. Previous researchers have introduced a correlation between extracted information based on the Hounsfield value of CBCT with other measured parameters related to dental implants. However, the accuracy of this method in defining density in three-dimensional has still not been tested, and the correlation between the Hounsfield value of CBCT and with true density of various objects is still unclear. Investigation to fill this gap is needed to gain the CBCT advantages in dental implants.

2.4 Procedures of Implant Placement

A dental implant that is placed to serve the structural support for a dental prosthesis has three main components: the implant, the abutment, and the prosthesis. The technique to install those implants is considered case by case; it might be a two-stage surgical procedure or a one-stage technique depending on clinical conditions and situations.

A two-stage surgical procedure is performed to place an implant in the jawbone by considering an adequate bone to support those dental implants. There are two phases of surgery in this procedure; in the first phase, the gingival tissue is opened, and the implant is placed in the proper site and covered back by gingival tissue. After osseointegration is achieved and the implant is fixed, the abutment is connected by reopening the gingival tissue and an artificial tooth is placed on the top of fixture (Natali 2003).

The one-stage technique is performed by placing the implant into the jawbone with only one surgical intervention. The abutment is placed directly into the implant before the implant is covered by gingival tissue. The difference technique is between two-stage and one-stage techniques.

Although the one-stage technique requires only a single surgical intervention, there are certain situations where the two-stage approach is more advantageous (Heijdenrijk 2000). This method allows for better protection of the bone from exposure, promoting faster bone regeneration. Additionally, it minimizes undesirable loading on the implant, which could affect stability during the critical osseointegration period. The two-stage technique also reduces the risk of infection during the healing process, particularly for patients unable to maintain proper oral hygiene. Moreover, it is often a more effective choice for patients scheduled to receive radiotherapy in the implant region, as it helps ensure a more predictable outcome.

2.5 Methods of Implant Stability Measurement in Dentistry

Implant stability is an indirect indication of osseointegration and is a measure of the clinical immobility of an implant (Parithimarkalaignan and Padmanabhan 2013). Implant stability is divided into two types: primary implant stability and secondary implant stability. Primary implant stability is defined as stability during implant insertion; meanwhile, secondary implant stability is defined as stability of dental implant after implant insertion. Implant stability is influenced by internal and external factors, including both material and local tissue-dependent variables. External factors that affect implant stability are the material of the implant, the length and diameter of the implant, its design, the micro-morphology of the implant surface and the implant insertion technique. Meanwhile, the important determinants as internal factors are the quality and quantity of the bone and the osseointegration process. The greatest primary stability of dental implants can be reached with simple drilling. The use of additional thread cutters and bone condensers has been shown to lessen primary stability significantly (Rabel et al. 2007).

Many efforts have been performed to assess dental implant stability biomechanically such as pull-out and push-out measurement, cutting torque resistance analysis, reverse torque test, and percussion test. However, these are destructive methods that are available for preclinical uses only (Chang and Giannobile 2012; Swami et al. 2016). The non-destructive methods that give clinical value are radiographic (CT, CBCT) and RFA methods. However, the resolution and variability while examination still should be improved (Spin-Neto et al. 2013; Kim 2014)

2.6 Tensional Test Method

The tensional test is used to test the interfacial tensile strength that was originally measured by detaching the implant plate from the supporting bone (Chang et al. 2010; Sachdeva et al. 2016). This technique is a destructive technique that is impossible to be applied to the patient pull-out and push-out measurement method.

The pull-out test is the tests that involve a tensile load dislodging a post from post space, while the push-out test uses the tensile load to be replaced by a compressive load (Ahmadian et al. 2013). That method is used to investigate the healing capabilities within the bone-implant interface (Haïat et al. 2013; Kempen et al. 2009). Based on this mechanism, the push-out and pull-out tests are only applicable for non-threaded cylinder-type implants, whereas most clinically available fixtures are of threaded design (Seong et al. 2013). Both methods are destructive; hence, it was not practised in the clinical study.

Pull-out and push-out tests were conducted on an in vitro model to estimate the implant stability. The result showed that this technique is very sensitive to modifications of technical details; therefore, the result of different models cannot be compared directly (Sakoh et al. 2006). In addition, because of this destructive technique, hence finite element method becomes a great method to explain the behaviour of the tensile mechanism of dental implants that cannot be achieved by push-out and pull-out tests. Chen et al. (2013) used the FEA to simulate the push-out test on a base model using three parameters: the diameter of the pin, the specimen's thickness, and the elastic modulus of the intracanal filter. The resulting stress was analysed to calculate the bone strength.

2.6.1 Removal Torque Method

The removal torque refers to the torsional force necessary for unscrewing the fixture using a torque manometer calibrated in Newton-centimetres (N-cm) and was first investigated by Johansson et al. (1998). The removal torque test measures the shear properties of the implant–tissue interface. However, this method is categorized as a destructive method.

Koh et al. (2009) conducted removal torque by using in vivo studies on the rabbit tibia model; the result showed that at the microscopic level, the fracture after removal torque test occurred at the implant-bone interface. Bardyn et al. (2010) applied the FEA method to investigate the mechanism of removal torque. In their pilot study on the replica of an implant inserted into polyurethane and sheep bone, the simulation result showed that there is a high correlation between estimation by FEA and a direct measurement of removal torque.

2.6.2 Cutting Resistance or Insertion Torque Method

The cutting resistance refers to the energy required in cutting a unit volume of bone while the insertion torque occurs during the fixture tightening procedure (Chang et al. 2010). Both of these methods can be used to estimate the primary implant stability (Wagner and Ka 2016; Bayarchimeg et al. 2013). However, these methods can be applied only during surgery. It is not applicable for monitoring post-surgery.

Dagher et al. (2014) measured the insertion torque of an ex vivo study on a sheep model. They showed that the insertion torque could be used as an indication of primary implant stability. This measurement has been cross-checked with Resonance Frequency Analysis (RFA) and the result showed that there is a significant positive correlation between insertion torque with RFA result.

A numerical study using FEA to investigate the influence of insertion torque on the stress distribution around an immediate oral implant has been conducted by Atieh et al. (2012). Their results showed that the use of insertion torque during the placement of dental implant increases the stress on crestal bone, and it was correlated with the bone quality of the site implant.

2.6.3 Resonance Frequency Analysis (RFA) Method

Resonance Frequency Analysis (RFA) is a non-invasive and non-destructive method that measures the implant stability objectively and reliably for any stage of the treatment or at follow-up examinations (Konstantinović et al. 2015). This method was introduced by Meredith et al. (1997) by developing the device called Osstell®. The implant stability measurement by this method is based on the micro-movement of an implant in its site that is reflected by a frequency resonance.

In the RFA device, a sinusoidal wave with certain frequency ranges (5–15 kHz) is generated and passed into a transducer that is attached to a small cantilever beam and it attached to the implant or abutment. The received signal will be received by a piezo ceramics element, and it will be transferred into a frequency analyzer (Digholkar et al. 2014). The peak of amplitude defines the excitation which shows a resonance frequency.

There are some important factors during measurement using the RFA system: transducer design, contact between implant with surrounding bone, and the effective length above the marginal bone level (Sennerby and Meredith 2008). To have an optimal measurement result and avoid the effect of soft tissue, the SmartPeg™ of the RFA device needs to be located at least 1–3 mm from the marginal bone level or 3 mm above the soft tissue. The direction of that SmartPeg should be perpendicular to the object.

Measurement of implant stability by using the RFA method has some advantages such as the consistency of measurement as object-oriented and measurement can be performed any time on any condition such as for immediate implant or delayed implant placement. Studies showed that implant stability (both primary and secondary implant stability) could be associated with osseointegration (Kunnekel et al. 2011; Shokri and Daraeighadikolaei 2013).

Breakthrough on implant stability measurement using RFA explored more detail on determinants of the dental implant. Implant stability measured by RFA can be related to other parameters such as insertion torque, bone density, and cortical thickness (Má et al. 2011; Trisi et al. 2011; Wada et al. 2015). However, the relation between dynamic properties of implants such as stress distribution, which is generated during loading of implants with different grade implant stability still not investigated yet.

2.7 Dental Imaging Technology

Radiography is an important aspect of dental treatment because it provides the image in detail of the site implant which is important for evaluation purposes. Site implant evaluation includes the dimension and availability of site implant and bone quality and quantity. Some imaging technologies such as CT scan, OPG scan, CBCT scan, and DEXA have been used in dentistry. However, the common problem in radiographic imaging is the bargaining between resolution and doses of X-ray exposure or radiation. Higher resolution images usually need higher doses of X-ray which are contradictory to the health issue.

2.7.1 Computerized Tomography (CT) Scan

Computerized tomography (CT) is an imaging technique that shows human anatomy in cross-section and provides a three-dimensional dataset that can be used for image reconstruction and analysis in several planes or three-dimensional settings. As well as another X-ray technique, a detector records the X-ray beam that passes through the patient's body. The computer then processes this information to create the CT image. CT technology provides 3-D images that can be used as diagnostic tools and treatment planning including evaluation of the quality and quantity of site implant (Karatas and Toy 2014).

CT technology can provide high-quality images by reducing the superimposition of images of structures outside the area of interest and high contrast between different tissues. CT technology can detect the density difference between tissues even if only 1% (Patel et al. 2009). The image in the CT scan can be viewed as images in the axial, coronal, or sagittal planes or in any arbitrary plane depending on the diagnostic task.

Another advantage of CT scan is that this technique can solve the artefact problem caused by metallic material by reducing it with the program during image reconstruction (Joemai et al. 2012; Schabel et al. 2016; Anderson 2016). Using CT images, radiologists can analyse the curvature of the maxillary and mandibular arch.

CT image scanning is constructed by an element of the image that is called a voxel. Each voxel has a value that is referred to in Hounsfield units (HU) which represent the density of the object at that position. The different tissues are identified as different HU. The ranges of the HU are from -1000 for air to $+3000$ for enamel; hence, the bone classification based on CT image is possible. Sençimen et al. (2011) classified bone using Misch's classifications based on HU as: muscle: 35–70 HU; fibrous tissue: 60–90 HU, cartilage: 80–130 HU; bone: 150–1800 HU and characterized bone quality such as D1 bone: > 1250 HU; D2 bone: 750–1250 HU; D3 bone: 375–750 HU; D4 bone: < 375 HU).

CT scan is a powerful technique to produce the accuracy of images with high resolution. However, the application for continuous monitoring is still debatable because of the radiation exposure and cost issues (Ogbole 2010; Lin 2010).

2.7.2 Cone Beam Computer Tomography (CBCT)

In the last decade, the CBCT machine has become a favourite dental imaging technique. This machine uses a cone type of source and a 2-D digital array of detectors to scan 3-D volumes in one scanning. The cone beam scanners are facilitated to rotate around the object of the head and neck up to 360 degrees; hence, multi-slicing (from 150 to more than 600) can be acquired at one time (Rao et al. 2012). With this technology, the examination time of the CBCT machine becomes shorter and the distortion effect due to the internal movement of the patient during scanning can be reduced.

The quality image of CBCT is controlled by controlling the field of view (FOV). However, if FOV is set larger, the quality image will decrease, the contrast between different tissues will lower and will produce more scattering (artefact) (Scarfe and Farman 2008; Hwang et al. 2016; Sonya et al. 2016). The quality image of CBCT depends on the size of the voxel of the area detector which is not dependent on the thickness of the slicing (Rao et al. 2012).

The imaging concept of CBCT is like CT which uses an attenuation of a monochromatic X-ray beam traversed in the material. The path of X-ray on CBCT from source to detector is a conical geometry (Miracle and Mukherji 2009). To get the image from CBCT, there are two main phases: data acquisition which uses cone-shaped X-ray beam and image reconstruction using computerized tomography.

Compared with CT machines, scanning with CBCT has some advantages, such as CBCT has a high spatial and temporal resolution, rapid scan, and dose reduction compared with CT. Hence, CBCT is more convenient for dental implant monitoring rather than conventional CT (Mohan et al. 2011; Shah et al. 2014; John et al. 2016).

CBCT has been applied for the evaluation of bone density; this evaluation is based on the assumption that there is a correlation between the attenuation of the X-ray of CBCT with the density (Naitoh et al. 2009; Silva et al. 2012). Hence, the density can be estimated from the Hounsfield unit (HU) obtained from CBCT. However, the scaling of greyscale from CBCT is still not standardized; the range can vary from -1500 to $+3000$ (Katsumata et al. 2007). To make more reliable to CT scale, some corrections need to be performed (Nomura et al. 2013).

Some of the advantages and disadvantages of CBCT scanning in dental imaging are summarized in the following table (Table 2.1).

Table 2.1 Advantages and disadvantages of CBCT (Gupta 2016; Pascual and Morale 2015)

Advantages of CBCT	Disadvantages of CBCT
• Low radiation doses of X-ray • High image accuracy (the range is from 0.4 mm to 0.125 mm) • Rapid scan time • Transferring data into different formats is easier. • Image quality due to artefacts can be improved by applying an artefact reduction algorithm	• Produce some artefacts • Sensitive with patient movement (positioning) • Poor soft tissue contrast • The scale is not standardized

2.8 Biomechanical of Dental Implant System

The component of the dental implant system at least consists of the material of the dental implant, cortical bone, trabecular bone, periodontal ligament, and internal process or modelling/remodelling of bone. The biomechanical dental system is a combination of all characteristics of its components.

The materials used in medical devices for dental implants are metals and their alloys. This material has a high Young's modulus and yields stress which provides stiffness to the structure. Every material implant should have biocompatibility with the human bone, be resistant to corrosion, and must have high fatigue strength and fracture toughness. The common materials for dental implants are stainless steel, cobalt, titanium, and pure titanium with appropriate alloys added to increase the mechanical characteristics. Examples of this material are palladium (Ti-0.2Pd), nickel-molybdenum (Ti-0.3Mo-0.8Ni) and aluminium and vanadium (Ti-6Al-4V). Not only will improve corrosion resistance but also with alloy addition will also increase the good shape-ability (Elias et al. 2008; Jogaib et al. 2015; Shemtov-Yona and Rittel 2015).

For components of the crown, ceramic porcelain is usually used. The hardness of this material is very high but more brittle. The advantage of this material is its colour that is close to natural teeth, and hence this material is more favourable to be used. Some material such as gold or zirconia is added into pure porcelain to gain the properties of porcelain material. The impurity of zirconia is more favourable to be used because this material has a similar colour to natural teeth and is also stronger than gold (Wassell, Walls and Steele 2002).

The main component of the implant dental system is the jawbone itself, which consists of cortical, and trabecular bone. Cortical bone is the hardest part of bone and covers all the surface of the bone (see Fig. 2.2). In the dental implant system, the cortical bone is the main support for sustaining the implant to be more stable. Hence, the elastic properties/stiffness of cortical bone are a key factor affecting the biomechanical system of dental implants. The hardness of cortical bone is different from one location to others; for example, between buccal side and lingual side is

2.8 Biomechanical of Dental Implant System

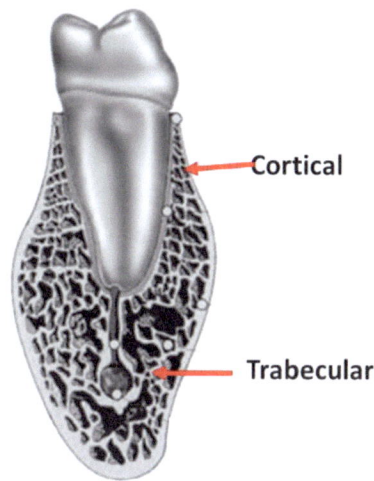

Fig. 2.2 Anatomy of bone: cortical and trabecular pictures (modified: Dent-Wiki.co http://www.dent-wiki.com/dental_technology/alveolar-bone-structure/)

different. The buccal side is harder than the lingual side. In other words, the properties of cortical bone are not homogenous/isotropy (O'Mahony et al. 2001).

Elastic properties of cortical bone are controlled by some factors such as hydration level, bone volume fraction, and bone mineral fraction. Increasing in bone mineral fraction of cortical bone will increase Young's modulus of bone. The relation between Young's modulus and bone volume fraction and bone mineral fraction has been reported by Wu et al. (2005).

Trabecular bone is a porous material called sponge bone or cancellous bone (Fig. 2.2). Like cortical bone, the trabecular bone also undergoes some internal processes such as modelling and remodelling. Modelling is an adaptation process of bone to change its shape as a response to mechanical loading. However, in adult modelling process is less than remodelling. Remodelling is an internal process of bone renewing and maintaining bone strength. The remodelling process includes the absorption of old bone, and it is replaced with new bone. The mechanical properties of bone are controlled by its bone mineral and organic matrix (Clarke 2008).

There is a relationship between the loading of dental implants with bone remodelling as reported and investigated by Degidi et al. (2005). They showed that bone remodelling was a significant difference in the immediate dental implant between unloaded and loaded. Loading can stimulate bone remodelling at the interface with significant differences compared with unloaded cases.

The loading process on dental implants is associated with daily mastication activity. Hence, the evaluation of biomechanicals of dental implants cannot be isolated from this process. Loading on the top of the implant or crown during mastication influences the dental implant mechanically. However, the biomechanical of the human mandible during masticatory is a complex process.

During mastication, forces are exerted on the dental implant in different directions. The stresses due to these forces will be transferred into adjacent teeth through the crown or enamel and periodontal ligament. Movement between each tooth also might be generated during mastication which depends on the anatomical surface of the proximal contacting surface and the elastic parameter of the object. The stresses generated with this process not only reach the neighbour teeth but also can reach the alveolar bone and make some deformation on it (Pileicikiene and Surna 2004).

2.9 Application of Finite Element Analysis for Dental Implant Assessment

The Finite Element Analysis (FEA) is a solution technique to solve a complex mechanics problem by dividing it into a set of small parts (elements) which has their own simple solution (Geng et al. 2008). In other words, FEA will solve the solution in each element instead of solving for the entire body and then combining each solution as a solution for the entire body. Fragmentation of the entire body into small parts which is called as meshing is needed before running FEA. This process is called discretization. In this process also, the element and their respective nodes and their boundary conditions are defined.

To solve the FEA equation, a matrix method is commonly used. The linear relations between displacements with forces/actions are re-arranged in the matrix form. There are several approaches to solving the FEA problem such as the flexibility approach and stiffness approach. In the flexibility approach, the relation between action (A) and displacement (D) is stated as follows:

$$D_1 = f_{11}A_1 + f_{12}A_2 + f_{13}A_3 + f_{14}A_4 \tag{2.1}$$

where f is a flexibility coefficient. Solving Eq. (2.1) needs a big matrix, especially for complex problems; hence this approach is not convenient to use because of computation reasons. Re-arranged Eq. (2.1) by moving the displacement as parameters that want to be solved on the right side and force/action as known parameters into the left side, this equation can be written as:

$$A_1 = k_{11}D_1 + k_{12}D_2 + k_{13}D_3 + k_{14}D_4 \tag{2.2}$$

where k is a stiffness coefficient. Equation (2.2) is called a stiffness approach, where the stiffness coefficient can be correlated directly with other elastic properties such as Young's modulus.

Dental implants are a complex system, and the application of finite element analysis needs some assumptions and simplifications. Those assumptions and simplifications include:

a. Smoothing of the object, for meshing purposes, the complex morphology structure of the real object will be simplified by the smooth object. Therefore, the detail of the complex real object cannot be achieved with this assumption.

2.9 Application of Finite Element Analysis for Dental Implant Assessment

b. Interfacing between surface contact area, because of complexity in the contact interface such as how to get the physical properties on the surface with the friction coefficient and size of the space that allowed objects to move to each other are difficult to define. Some simplifications are conducted such as interfacing between implant and jawbone is assumed coupled perfectly. It means that between implant and jawbone is assumed there is no space to make some slip or friction or micro-movement, which is far from the real condition where the space between implant and jawbone, is possibly laminated by thin periodontal ligament (PDL) (Papavasiliou et al. 1997).

c. Material properties assignment, it is difficult to have a model with material properties exactly like the real object where the properties of real object are non-homogenous. To simplify the problem, usually, a homogenous model of material is used. The heterogeneity and anisotropy, which represent the variation of density and portion of cortical-trabecular bone, are ignored (Chirchir 2016).

Because the solution of FEA resulted from a series of approximations and limitations, the interpretation of this needs to be aware of some factors such as the definition of material properties, loading schemes and time-based responses, and contact behaviours. Hence, during the interpretation of the results, the user needs to be aware of that limitation.

FEA technique has been applied widely in dentistry especially to understand the dynamic properties of the system such as stress, strain, and micromotion of dental implants. Lin et al. (2007a) used FEA to quantify the stress and strain developed in the bony tissues and compared the stress and strain progressions at different healing stages. They tried to investigate the effect of material properties changes on bone due to the bone remodelling process through a 3-D finite element model, which is derived from 3-D CT scanning.

Based on the FEA study, many parameters of dental implants can be identified in their relationship with generated dynamic processes such as stress and strain distribution; the geometry of the implant is known to have an impact on biomechanical responses, and the generated in the jawbone can be identified. Lin li et al. (2005) investigated the influence of dental implant length and bone quality on biomechanical response in bone around implant using nonlinear FEA.

Chaichanasiri et al. (2009) used FEA to investigate the influence of premature contact caused by an implant-retained crown (IRC) on stress and strain distributions in the bone surrounding the implant. Their result showed that the magnitude of Von Mises stresses (the critical stress when the material is at the failure condition) in the bone changes drastically when there is premature contact. In other words, Von Mises stress is a value used to determine if a given material will yield or fracture. It is mostly used for ductile materials, such as metals. The Von Mises yield criterion states that if the Von Mises stress of a material under load is equal or greater than the yield limit of the same material under simple tension which is easy to determine experimentally, then the material will yield. (https://www.simscale.com).

References

L. Ahmadian, R. Arbabi, J. Kashani, Original research compression of stress distribution in pull out and push out bond strength test set ups: a 3-D finite element stress analysis. Int. J. Prosthetic Dent. **4**(2231), 1–8 (2013)

L.H. Al-Nakib, Computed tomography bone density in Hounsfield units at dental implant receiving sites in different regions of the jaw bone. J. Baghdad College Dent. **26**(March), 92–97 (2014)

S.E. Anderson, Treatment planning on CT images reconstructed with an iterative metal artifact reduction algorithm. Med. Phys. Radiat. Oncol. 1–32 (2016)

P. Aranyarachkul, J. Caruso, B. Gantes, E. Schulz, M. Riggs, I. Dus, J.M. Yamada, M. Crigger, Bone density assessments of dental implant sites: 2. Quantitative cone-beam computerized tomography. Int J. Oral Maxillofac Implants. **20**(3), 416–24 (2005). PMID: 15973953

M.A. Atieh, N.H.M. Alsabeeha, A.G.T. Payne, D.R. Schwass, D. Pros, W.J. Duncan, F. Perio, Insertion torque of immediate wide-diameter implants: a finite element analysis. Quintessence Int. **43**(9), 115–126 (2012)

T. Bardyn, P. Gédet, W. Hallermann, P. Buchler, Prediction of dental implant torque with a fast and automatic finite element analysis: a pilot study. Oral Maxillofac. Radiol. **109**(4), 594–603 (2010)

D. Bayarchimeg, H. Namgoong, B.K. Kim, M.D. Kim, S. Kim, T. Kim, K. Koo, Evaluation of the correlation between insertion torque and primary stability of dental implants using a block bone test. J. Periodontal Implant Sci. **43**, 30–36 (2013)

E1. Chaichanasiri, P. Nanakorn, W. Tharanon, J.V. Sloten, Finite element analysis of bone around a dental implant supporting a crown with a premature contact. J. Med. Assoc. Thai. **10**, 1336–1344 (2009)

P.C. Chang, W.V. Giannobile, Functional assessment of dental implant osseointegration. Int. J. Periodont. Restorat. Dent. **32**(5), e147–e153 (2012)

P.-C. Chang, N.P. Lang, W.V. Giannobile, Evaluation of functional dynamics during osseointegration and regeneration associated with oral implants: a review. Clin. Oral Investigat. **16**(3), 679–688 (2010)

H.-W. Chang, H.-L. Huang, J.-H. Yu, J.-T. Hsu, Y.-F. Li, Y.-F. Wu, Effects of orthodontic tooth movement on alveolar bone density. Clin. Oral. Investig. **16**(3), 679–88 (2012)

W. Chen, Y. Chen, S. Huang, C. Lin, Limitations of push-out test in bond strength measurement. J. Endod. **39**(2), 283–287 (2013)

H. Chirchir, Limited trabecular bone density heterogeneity in the human skeleton. Anatom. Res. Int. (2016)

P.M. Chopra, Correlation of cone beam computerized tomography bone density with cutting resistance values as obtained at the time of implant placement. Res. Graduat. Stud. Tex. (2010)

B. Clarke, Normal bone anatomy and physiology. Clin. J. Am. Soc. Nephrol. **3**, 131–139 (2008)

M. Dagher, N. Mokbel, J. Gabriel, N. Naaman, Resonance frequency analysis, insertion torque, and bone to implant contact of 4 implant surface: comparison and correlation study in sheep. Implant Dentistr. 1–7 (2014)

M. Degidi, A. Scarano, M. Piattelli, V. Perrotti, A. Piattelli, Bone Remodeling in immediately loaded and unloaded titanium dental implants: a histologic and histomorphometric study in humans. J. Oral Implantol. **31**(1), 18–24 (2005)

S. Digholkar, V. Naga, V. Madhav, J. Palaskar, S. Digholkar, A. Statistics, E. Alert, Methods to measure stability of dental implants. J. Dent. Allied Sci. **3**(1), 17–23 (2014)

C.N. Elias, J.H.C. Lima, R. Valiev, M.A. Meyers, Biomedical applications of titanium and its alloys. J. Minerals Metals Mater. Soc. (March), 46–49 (2008)

M. Fanuscu, T. Chang, Three-dimensional morphometric analysis of human cadaver bone: microstructural data from maxilla and mandible. Clin. Oral Implants Res. **15**(2), 213–218 (2004)

F. Faot, M. Chatterjee, G.V. de Camargos, J. Duyck, K. Vandamme, Micro-CT analysis of the rodent jaw bone micro-architecture: a systematic review. Sci. Direct Bone Rep. **2**, 14–24 (2015)

References

B. Friberg, L. Sennerby, J. Roos, U. Lekholm, Identification of bone quality in conjunction with insertion of titanium implants. A pilot study in jaw autopsy specimens. *Clinical Oral Implants Research,* 6(4), *213–9* (1995)

H. Fuller, R. Fuller, R.M.R. Pereira, High resolution peripheral quantitative computed tomography for the assessment of morphological and mechanical bone parameters. Rev. Bras. Reumatol. **55**(4), 352–362 (2015)

J. Geng, W. Yan, W. Xu, K.B.C. Tan, S. Lee, H. Xu, J. Chen, Finite element modelling in implant dentistry, in *Advanced Topics in Science and Technology in China* (2008)

D. Gupta, *The role of CBCT Imaging in Dentistry by. Dentaltown Magazine, (March)* (2016)

G. Haïat, H.-L. Wang, J. Brunski, Effects of biomechanical properties of the bone-implant interface on dental implant stability: from in silico approaches to the patient's mouth. Ann. Rev. Biomed. Eng. **16**, 187–213 (2013)

K. Heijdenrijk, Two-stage dental implants inserted in a one-stage procedure. J. Clin. Periodontol. **29**(10), 901–909 (2000)

J.J. Hwang, H. Park, H.G. Jeong, S.S. Han, Change in image quality according to the 3D locations of α CBCT phantom. PLoS ONE **11**(4) (2016)

K. Isoda, Y. Ayukawa, Y. Tsukiyama, M. Sogo, Y. Matsushita, K. Koyano, Relationship between the bone density estimated by cone-beam computed tomography and the primary stability of dental implants. Clin. Oral Implants Res. **23**(7), 832–836 (2012)

R.M.S. Joemai, P.W. de Bruin, W.J.H. Veldkamp, J. Geleijns, Metal artifact reduction for CT: development, implementation, and clinical comparison of a generic and a scanner-specific technique. Med. Phys. **39**(January), 1125 (2012)

D. Jogaib, C. Nelson, R. Zufarovich, Properties and performance of ultrafine grained titanium for biomedical applications. Mater. Res. **18**(6), 1163–1175 (2015)

C.B. Johansson, C.H. Han, A. Wennerberg, T. Albrektsson, A quantitative comparison of machined commercially pure titanium and titanium-aluminum-vanadium implants in rabbit bone. Int. J. Oral Maxillof. Implants **13**, 315–21 (1998)

G.P. John, T.E. Joy, J. Mathew, V.R.B. Kumar, Applications of cone beam computed tomography for a prosthodontist. J. Ind. Prosthodont. Soc. **16**(1), 3–7 (2016)

O.H. Karatas, E. Toy, Three-dimensional imaging techniques: a literature review. Eur. J. Dentistr. **8**(1), 132–140 (2014)

A. Katsumata, A. Hirukawa, S. Okumura, M. Naitoh, M. Fujishita, E. Ariji, R.P. Langlais, Effects of image artifacts on gray-value density in limited-volume cone-beam computerized tomography. Oral Surg. Oral Med. Oral Pathol. Oral Radiol. Endodontol. **104**(6), 829–836 (2007)

D.H.R. Kempen, L. Lu, A. Heijink, T.E. Hefferan, L.B. Creemers, A. Maran, W.J. Dhert et al., Effect of local sequential VEGF and BMP-2 delivery on ectopic and orthotopic bone regeneration. Biomaterials **30**, 2816–2825 (2009)

P.G. Khayat, H.M. Arnal, B.I. Tourbah, L. Sennerby, Clinical outcome of dental implants placed with high insertion torques (Up to 176Ncm). Clin. Implant Dent. Relat. Res. **15**, 227–233 (2013)

D.-G. Kim, Can dental cone beam computed tomography assess bone mineral density? J. Bone Metabol. **21**(2), 117–126 (2014)

U. Kini, B.N. Nandeesh, Radionuclide and Hybrid Bone Imaging, in *Radionuclide and Hybrid Bone Imaging* (2012), pp. 29–57

J. Koh, J. Yang, J. Han, J. Lee, S. Kim, Biomechanical evaluation of dental implants with different surfaces: removal torque and resonance frequency analysis in rabbits. J. Adv. Prosthodont. **1**, 107–112 (2009)

V.S. Konstantinović, F. Ivanjac, V. Lazić, I. Djordjević, Assessment of implant stability by resonant frequency analysis. Vojnosanit. Pregl. **72**(2), 169–174 (2015)

A.T. Kunnekel, M.T. Dudani, C.K. Nair, E.M. Naidu, G. Sivagami, Comparison of delayed implant placement vs immediate implant placement using resonance frequency analysis: a pilot study on rabbits. J. Oral Implantol. **37**(5), 543–548 (2011)

U. Lekholm, G.A. Zarb, Patient selection and preparation, in *Tissue Integrated Prostheses: Osseointegration in Clinical Dentistry* (Quintessence Publishing, 1985)

G. Lester, Bone quality: summary of NIH/ASBMR meeting. J. Musculosk. Neuronal Interact. **5**(4), 309 (2005)

E.C. Lin, Radiation risk from medical imaging. Mayo Clin. Proc. **85**(12), 1142–1146 (2010)

C.L. Lin, Y.-C. Kuo, T.S. Lin, Effects of dental implant length and bone quality on biomechanical responses in bone around implants: a 3-D non-linear finite element analysis. Biomed. Eng. Appl. Basis Commun. **17**, 44 (2005)

D. Lin, Q. Li, W. Li, I. Ichim, M. Swain, Biomechanical evaluation of the effect of bone remodeling on dental implantation using finite element analysis, in *5th Australasian Congress on Applied Mechanics, ACAM 2007*, 10–12 December 2007, Brisbane, Australia (2007a)

L.I. Linkow, *Theories and Techniques of Oral Implantology* (2012), pp. 2–3

F. Má, P.M. Diago, P.D. Oltra, Relationships between bone density values from cone beam computed tomography, maximum insertion torque, and resonance frequency analysis at implant placement: a pilot study Commons. Int. J. Oral Maxillofac. Implants **26**(5), 1051–1056 (2011)

M. Marquezan, T.C.L. Lau, C.T. Mattos, Da A.C. Cunha, L.I. Nojima, E.F. Sant'Anna, M.T.D.S. Araújo, Bone mineral density Methods of measurement and its influence on primary stability of miniscrews. Angle Orthodont. **82**(1), 62–6 (2012)

N. Meredith, D. Science, L. Maudlin, Resonance frequency measurements of implant stability in viva. A cross-sectional and longitudinal study of resonance frequency measurements on implants in the edentulous and partially dentate maxilla., 2–3. Clin. Oral. Impl. Res. **8**, 226–233 (1997)

C. Miracle, S.K. Mukherji, Conebeam CT of the head and neck, part 1: physical principles. AJNR. Am. J. Neuroradiol. **30**(6), 1088–95 (2009)

C.E. Misch, H.A. Abbas, *Contemporary Implant Dentistry*, 3rd ed. (Mosby, St. Louis, 2008), pp. 38–64; 130–144; 645–665 (2008)

R. Mohan, A. Singh, M. Gundappa, Three-dimensional imaging in periodontal diagnosis—utilization of cone beam computed tomography. J. Ind. Soc. Periodontol. **15**(1), 11–17 (2011)

M. Naitoh, A. Hirukawa, A. Katsumata, E. Ariji, Evaluation of voxel values in mandibular cancellous bone: relationship between cone-beam computed tomography and multislice helical computed tomography. Clin. Oral Implants Res. **20**(5), 503–506 (2009)

A. Natali, *Dental Biomechanics* (Taylor and Francis Inc, USA and Canada, 2003)

Y. Nomura, H. Watanabe, K. Shirotsu, E. Honda, Y. Sumi, T. Kurabayshi, Stability of voxel values from cone-beam computed tomography for dental use in evaluating bone mineral content. Clin. Oral Implants Res. **24**(5), 543–548 (2013)

M.R. Norton, C. Gamble, Bone classification: an objective scale of bone density using the computerized tomography scan. Clin. Oral Impl. Res. **12**, 79–84 (2001)

M. O'Mahony, J.L. Williams, P. Spencer, Anisotropic elasticity of cortical and cancellous bone in the posterior mandible increases peri-implant stress and strain under oblique loading. Clin. Oral Implants Res. **12**(6), 648–657 (2001)

G.I. Ogbole, Radiation dose in paediatric computed tomography: risks and benefits. Ann. Ibadan Postgraduat. Med. **8**(2), 118–126 (2010)

G. Papavasiliou, P. Kamposiora, S.C. Baynet, D.A. Felton, 3D-FEA of osseointegration percentages and patterns on implant-bone interfacial stresses. J. Dentistry **25**(6), 485–491 (1997)

S. Parithimarkalaignan, T.V. Padmanabhan, Osseointegration: an update. J. Ind. Prosthod. Soc. **13**(1), 2–6 (2013)

J. Pascual, K. Morale, ScienceDirect Endodontic applications of cone beam computed tomography: case series and literature review. ScienceDirect, 38–50 (2015)

S. Patel, A. Dawood, E. Whaites, T. Pitt Ford, New dimensions in endodontic imaging: Part 1. Conventional and alternative radiographic systems. Int. Endodontic J. **42**, 447–462 (2009)

G. Pileicikiene, A. Surna, The human masticatory system from a biomechanical perspective: a review. Baltic Dent. Maxillofac. J. **6**, 81–84 (2004)

A. Rabel, S.G. Kohler, A.M.S. Westhausen, Clinical study on the primary stability of two dental implant systems with resonance frequency analysis. Clin. Oral Invest. **11**(3), 257–265 (2007)

References

G.V. Rao, S.A. Rao, P.M. Mahalakshmi, E. Soujanya, Cone beam computed tomography—an insight beyond eyesight in clinical dentistry. Innovative J. Med. Health Sci. **2**, 74–80 (2012)

A. Sachdeva, P. Dhawan, S. Sindwani, I.M. El-Sayed El-Hakim, Assessment of implant stability: methods and recent advances. Br. J. Med. Med. Res. Critic. Care Trauma Hosp. India **12**(3), 1–10 (2016)

J. Sakoh, U. Wahlmann, E. Stender, B. Al-Nawas, W. Wagner, Primary stability of a conical implant and a hybrid, cylindric screw-type implant in vitro. Int. J. Oral Maxillofac. Implants **21**(4) (2006)

W.C. Scarfe, A.G. Farman, What is cone-beam CT and how does it work? Dent. Clin. North Am. **52**(4), 707–730 (2008)

C. Schabel, S. Gatidis, M. Bongers, F. Hüttig, G. Bier, J. Kupferschlaeger, C. Pfannenberg, Improving CT-based PET attenuation correction in the vicinity of metal implants by an iterative metal artifact reduction algorithm of CT data and its comparison to dual-energy–based strategies. Invest. Radiol. 1 (2016)

M. Sençimen, A. Gülses, J. Ozen, C. Dergin, K.M. Okçu, S. Ayyıldız, H.A. Altuğ, Early detection of alterations in the resonance frequency assessment of oral implant stability on various bone types: a clinical study. J. Oral Implantol. **37**, 411–419 (2011)

L. Sennerby, N. Meredith, Implant stability measurements using resonance frequency analysis: biological and biomechanical aspects and clinical implications. Periodontol. **2000**(47), 51–66 (2008)

W.J. Seong, S. Grami, S.C. Jeong, H.J. Conrad, J.S. Hodges, Comparison of push-in versus pull-out tests on bone-implant interfaces of rabbit tibia dental implant healing model. Clin. Implant Dent. Relat. Res. **15**, 460–469 (2013)

N. Shah, N. Bansal, A. Logani, Recent advances in imaging technologies in dentistry. World J. Radiol. **6**(10), 794–807 (2014)

T. Shapurian, P.D. Damoulis, G.M. Reiser, T.J. Griffin, W.M. Rand, Quantitative evaluation of bone density using the Hounsfield index. Int. J. Oral & Maxillofac. Implants **21**(2), 290–297 (2006)

K. Shemtov-Yona, D. Rittel, An overview of the mechanical integrity of dental implants. BioMed. Res. Int. 547384 (2015)

M. Shokri, A. Daraeighadikolaei, Measurement of primary and secondary stability of dental implants by resonance frequency analysis method in mandible. Int. J. Dentistr. 1–6 (2013)

H. Sievänen, P. Kannus, T.L.N. Järvinen, Bone quality: an empty term. PLoS Med. **4**(3), e27 (2007)

I.M. Silva, D.Q. Freitas, G.M. Ambrosano, F.N. Bóscolo, S.M. Almeida, Bone density: comparative evaluation of Hounsfield units in multislice and cone-beam computed tomography. Brazil. Oral Res. **26**(6), 550–6 (2012)

D. Sonya, J. Davies, N. Ford, A comparison of cone-beam computed tomography image quality obtained in phantoms with different fields of view, voxel size, and angular rotation for iCAT NG. J. Oral Maxillofac. Radiol. **4**(2), 31–39 (2016)

R. Spin-Neto, E. Gotfredsen, A. Wenzel, Impact of voxel size variation on CBCT-based diagnostic outcome in dentistry: a systematic review. J. Digit. Imag. **26**(4), 813–820 (2013)

V. Swami, V. Vijayaraghavan, V. Swami, Current trends to measure implant stability. J. Indian Prosthodont. Soc. **16**, 124–130 (2016)

P. Trisi, M. Todisco, U. Consolo, D. Travaglini, High versus low implant insertion torque: a histologic, histomorphometric, and biomechanical study in the sheep mandible. Int. J. Oral Maxillofac. Implants **26**(4), 837–849 (2011)

I. Turkyilmaz, E.A. McGlumphy, Influence of bone density on implant stability parameters and implant success: a retrospective clinical study. BMC Oral Health **8**, 32 (2008)

M. Wada, Y. Tsuiki, T. Suganami, K. Ikebe, M. Sogo, I. Okuno, The relationship between the bone characters obtained by CBCT and primary stability of the implants. Int. J. Implant Dentistr. **1**(2), 1–7 (2015)

W. Wagner, G. Ka, Insertion torque and resonance frequency analysis of dental implant systems in an animal model with loaded implants. Int. J. Oral Maxillofac. Implants **21**(5), 726–732 (2016)

R.W. Wassell, A.W.G. Walls, J.G. Steele, Crowns and extra-coronal restorations: materials selection. Br. Dental J. **192**(4), 111–199 (2002)

J.S. Wu, H. Lin, J.-P. Hung, J.-H. Chen, Effects of bone mineral fraction and volume fraction on the mechanical properties of cortical bone. J. Med. Biol. Eng. **26**(1), 1–7 (2005)

Chapter 3
Digital Imaging and Implant Stability, and Finite Element Analysis of Study Preparation

3.1 Introduction

To achieve all the research objectives, the clinical treatments that involve dental implant patients and a series of laboratory work have been conducted. This research deals with in vitro and in vivo data. Therefore, ethical approval is needed before starting the research. The ethical approval was obtained from Hospital USM (Hospital USM). All the research activity related to the patient treatments is conducted by an oral surgeon from Hospital USM; the researcher is not involved directly in those clinical treatments of the patients.

3.2 Ethical Approval

Ethical approval for this study was obtained from the Human Research and Ethics Committee of University Sains Malaysia (USM) [No.254.4(1.3)]. The title of the approved study was Biomechanics of Dental Implant: Three-Dimensional Bone Assessment Using Cone Beam Computer Tomography (CBCT) and Dual Energy X-Ray Absorptiometry (DEXA) from Laboratory to Clinic. However, because of the unavailability of instruments, the use of DEXA is eliminated.

3.2.1 Study Design

In general, this research is divided into two types of studies that are in vitro and in vivo studies. The schematic of all work is shown in Fig. 3.1.

In the in vivo study, data are collected in two ways: observational perspective and longitudinal design. Observational perspective studies are conducted to

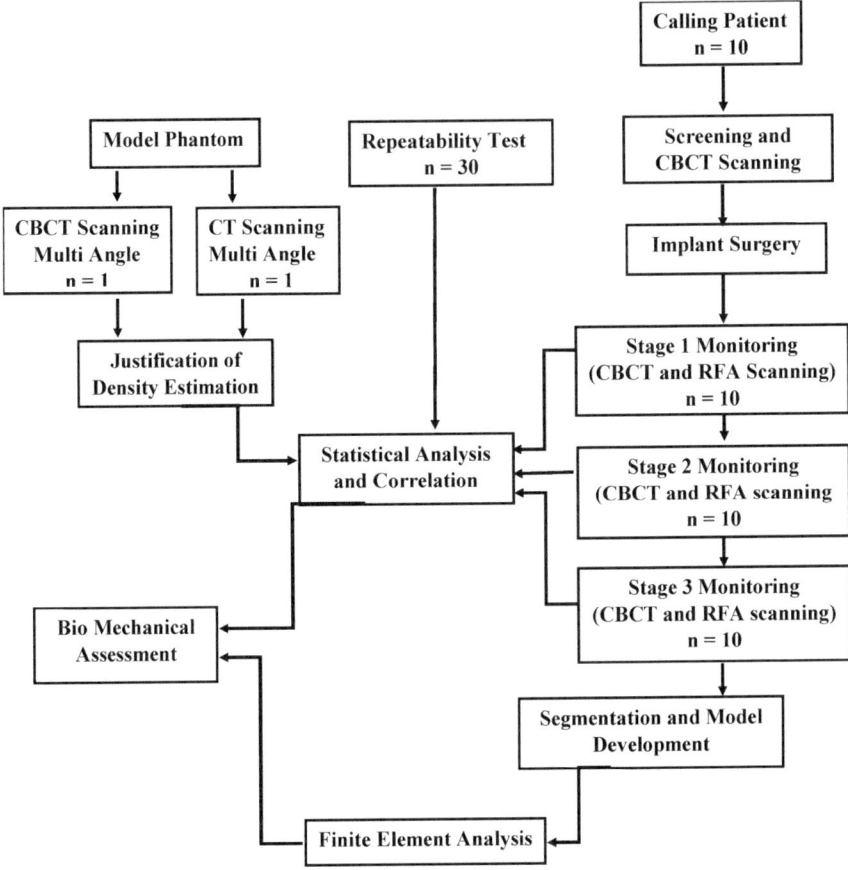

Fig. 3.1 Workflow of research consists of in vivo and in vitro studies

measure bone density which is associated with the osseointegration process during implant treatment of the patient, as mentioned in objective 3. In this study, 10 dental implant patients are included for statistical analysis purposes. The direct treatment of dental implants, including evaluation of CBCT to assess the availability of implant, implant surgery, cast impression, crown installation, and evaluation post-implant, is performed by a professional at Hospital USM using the standard procedure at Hospital USM. In this research, every patient is treated with the same implant type that is MegaGen Implant Co., Ltd., South Korea, with the size adjusted based on the patient's conditions.

Longitudinal studies are performed to investigate the relationship between secondary implant stability and the quality of bone as stated in objective 2. The same samples of the patients as well as in cross-sectional study are involved. The assessment of implant stability was conducted in three phases. The first assessment is conducted on the same day as implant placement (immediately after implant surgery), and it is called stage 1. The data collection during this stage includes the CBCT scanning and Resonance Frequency Analysis (RFA) measurement. The second stage (so-called stage 2) for the assessment of the implant stability is performed about 3 months after implant placement; this condition represents the preloading condition. The third stage of assessment (stage 3) is performed 4 months after implant placement or 1 month after crown installation; this condition represents the post-loading condition. For each stage assessment, CBCT scanning and RFA are performed to measure the density and implant stability in ISQ value. The density through CBCT measurements is performed to monitor the remodelling and osseointegration of the bone surrounding the implant during implant treatment.

3.2.2 Reference Population

All subjects for dental implant placement undergoing CBCT were obtained from the Hospital USM. The inclusion criteria for implant patient selection are age between 25 and 55 years old, willing to do the program, natural single tooth edentulous on mandible case, the patient who has edentulous at molar teeth. Meanwhile, the exclusion criteria are a smoker and a person who has no systemic disease such as heart problems, high blood pressure, and diabetes.

3.2.3 Sample Size

The number of sample sizes which are included in this research is determined using StatsDirect statistical software V 2.7.8. In this calculation, the sample size is derived from the Pearson correlation between CBCT and RFA methods that resulted from previous researchers. The inputs for this calculation are Power = 90%, ALPHA = 5%, $RH0 = 0$, Alt $RH = 0.56$; hence the estimated sample size is 10 patients.

3.2.4 Statistical Analysis

The Statistical Package of Social Science (SPSS) version 18.0 is used for data entry and statistical analysis. Paired t-test has been used to compare the two group measurements of pre- and post-implant using either CBCT or RFA. The FEA result from in vitro study has been compared with the direct measurement of RFA using an unpaired t-test.

3.3 Procedure of Implant Placement

In this study, the procedure of implant that is used was following the Hospital USM procedure by using a two-stage implantation technique. On the day of surgery, the patient is administered topical and local anaesthesia followed by implant surgery and implant installation. For dental implant placement, the dental submerged; this placement of implant the sub-merger type where healing screw would close (suture) resorbable about 3 mm. The illustrations of implant placement for molar edentulous are shown in Fig. 3.2.

The next step during implant insertion is healing screw placement. The healing screw is placed to protect the exposed portion of the dental implant; hence, it can seal the internal part of the dental from the oral environment. The gum tissue then is sutured to cover the area of the implant and healing screw or cap. The healing screw, condition after healing screw insertion, and suture condition are shown in Fig. 3.3.

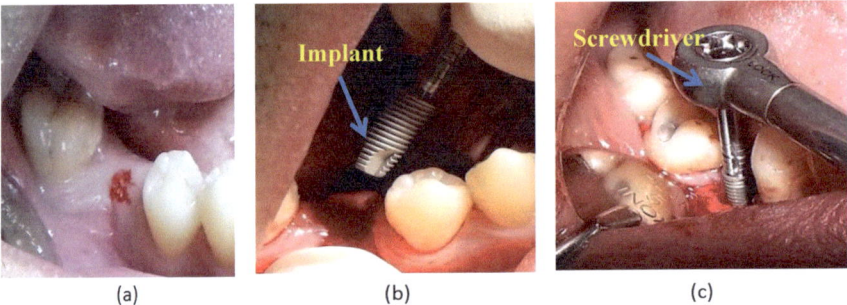

Fig. 3.2 **a** Loss of the molar tooth, **b** placement implant insertions, and **c** torquing wrench/screwdriver instrument is being used to tighten the implant

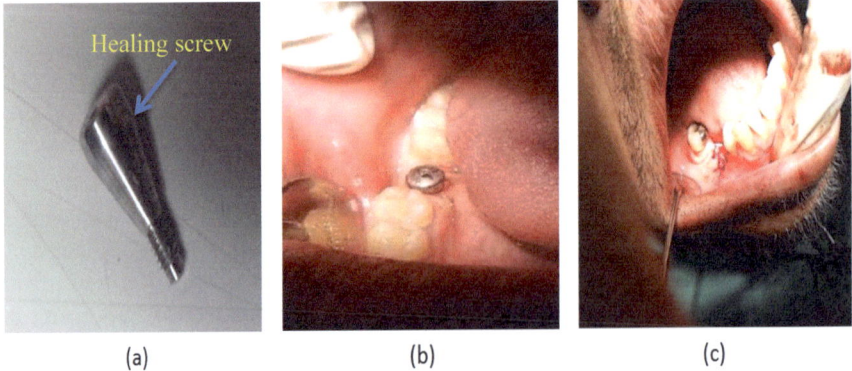

Fig. 3.3 **a** Healing screw, **b** condition after insertion of the healing screw, and **c** condition after suture

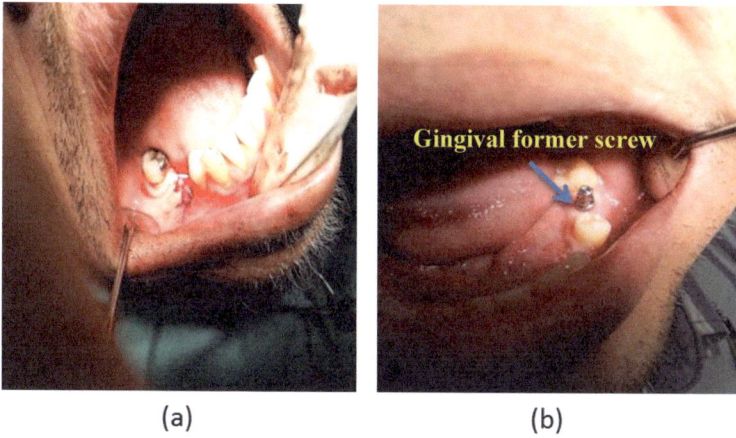

Fig. 3.4 a Suture after implant placement, and **b** after 3 months, replacement healing screw and placement of the gingival former screw

After 3 months, the patient is called to remove the healing screw and replacement with a gingival former screw. The impression that consists of the step removing the gingival former screw was performed after 1 week from placement, and it was replaced by transfers coping and proceed impression with polyvinyl siloxane (PVS). The transferred coping screw together with the impression tray is sent to a lab for crown construction and healing screw replacement at the implant site (Fig. 3.4).

Finally, then the implant is loaded immediately with a temporary screw-retained crown. The screw retain was used to make it easier for RFA measurement purposes after 1 month of crown placement.

3.4 Procedure of Dental Impression

The dental impression is taken 1 week after placement gingival screw former and unscrews the gingival screw for these purposes. Then the transfer coping is placed to replace the gingival screw. The dental impression was started by inserting the tray that was filled with alginate to cast the maxilla side and the patient was asked to hold his/her tooth several times until the material became a gel. The condition of the implant after removing the gingival and after replaced by transfer coping is shown in Fig. 3.5

After the tray for maxilla is finished and taken out it from mouth, the new tray for casting of mandible is inserted. The examic material is sprayed into the tray until it covered the whole space on the implant and surrounding implant. After the material becomes solid, the tray is taken out from the mouth, the transfers coping is put back, and it is placed into the mandible tray. The healing screw again is placed back to cover the implant. Illustration for maxilla casting and tray for the mandible that ready

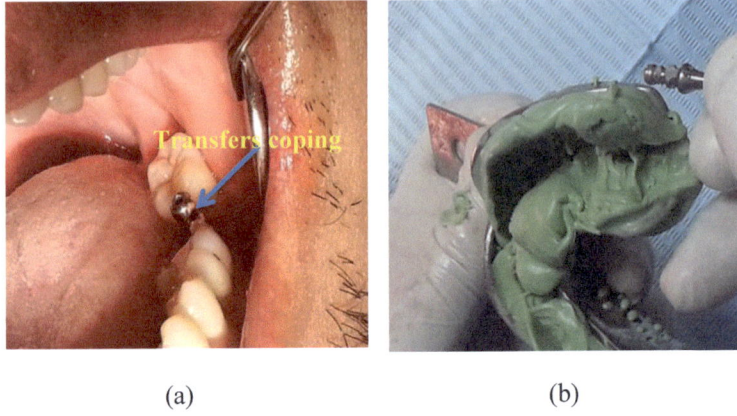

Fig. 3.5 **a** Condition of implant 1 week after placement gingival former screw, unscrew the gingival screw for placement the transfers coping into the implant, and **b** transfers coping with screw partly intruded

to be sent for next steps, and implant condition after installed back the healing screw cast is shown in Fig. 3.6.

All those trays are sent to the laboratory for crown fabrication. In the laboratory, crown fabrication started by filling the tray with moulded material to cast the maxilla and mandible teeth. In the site implant target was covered by plastic to simulate the gum tissue and to make it easier the remove the abutment after the crown fabrication. The next step is abutment installation on-site implant; the abutment is cut to make the length of the abutment and is same as an available space during biting conditions at this location. The mandibular model is cut and also follows the space available in the area between the teeth, and then the crown material is filled in the implant location to cast the edentulous crown. This crown model is trimmed following the morphology of the tooth then furnished and polished to make it more resemble the

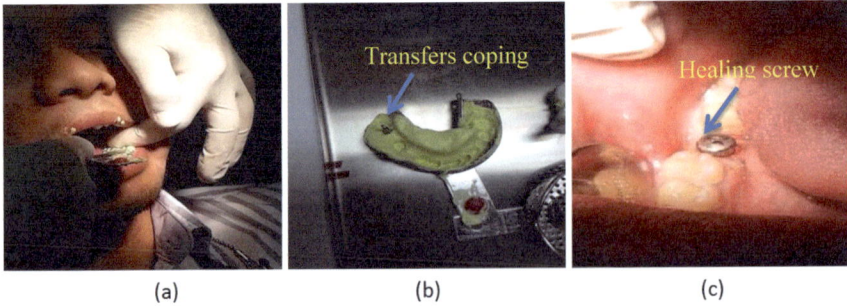

Fig. 3.6 **a** Dental impression will be used later when the dentist is ready to make the crown, **b** impression of the lower teeth with the transfers coping in place is now sent to the lab, and **c** screwing the healing cap back on to the implant for two weeks

3.5 Procedure of Crown Installation

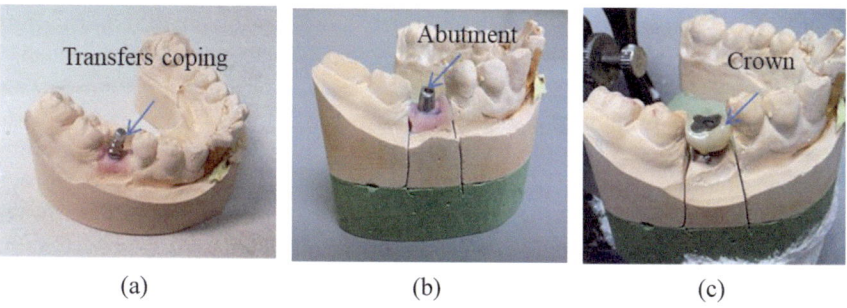

Fig. 3.7 Crown will be ready in 2 weeks. Before that crown installed into patient, the finishing, refinement, and repositioning with maxillaries and mandibular teeth are performed

natural tooth. The abutment before and after cutting installed in the cast impression tray and crown cast is shown in Fig. 3.7.

The crown will be ready in 2 weeks. Before that crown is installed into a patient, the finishing, refinement, and repositioning with maxillaries and mandibular teeth are performed.

3.5 Procedure of Crown Installation

After finishing the polishing and testing repositioning with the maxillaries and mandibular teeth model, the crown and abutment were lifted out from the tray cast impression model. The abutment and crown were ready to be installed into the patient, and the patient was called for crown installation (Fig. 3.8).

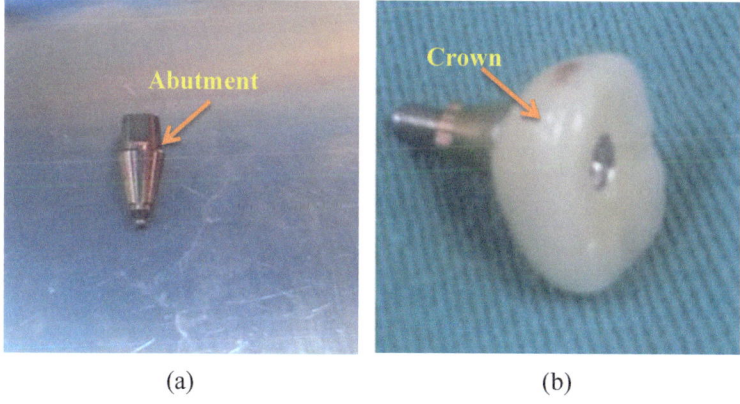

Fig. 3.8 **a** Abutment and **b** crown ready for installation

In the crown installation stage, the first step is unscrewing the healing cap from the implant and placing the abutment in the implant. The abutment is tightened until the maximum interlocking is achieved. The illustration of this step for unscrewing the healing cap and abutment installed is shown in Fig. 3.9.

After the abutment is installed properly in the implant, the next step is inserting the crown into the abutment. Before installation, some epoxy/glue was filled into the crown to make better interlocking between crowns with abutment. Finally, the last step in implant treatment is cementation of the surface of the crown to close the opened crown surface. The installed crown and after-cementation step are shown in Fig. 3.10. Then the patient asked to be scanned using CBCT for monitoring purposes (stage 3). Implant stability at post-crown condition is measured one month after crown installation.

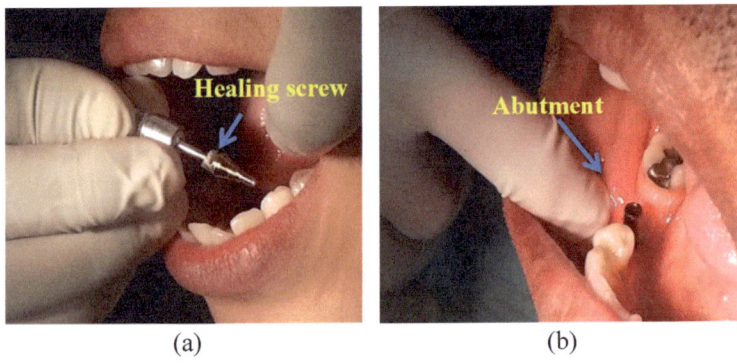

Fig. 3.9 **a** Unscrewing the healing cap from the implant and **b** installed abutment into the implant

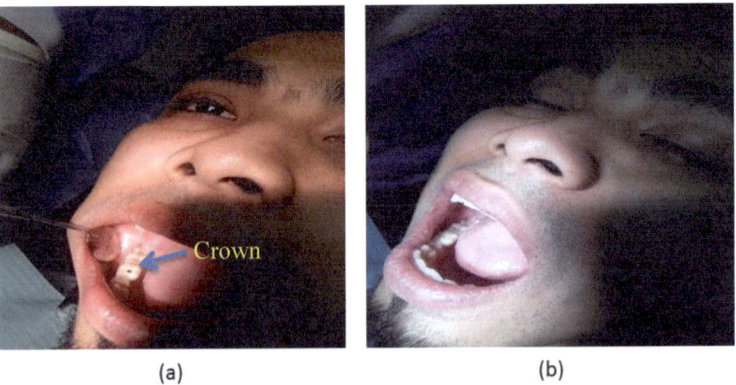

Fig. 3.10 **a** After the crown is installed into the abutment and **b** after cementation of the crown as the final stage of implant treatment

3.6 Dental Imaging and Implant Stability Measurement

The dental imaging during treatment is performed by using CBCT periodically following the stage of implant treatment. During the period of monitoring, not only CBCT scanning was performed but also the implant stability measurement. The implant stability was measured by using RFA. The schedule for CBCT scanning and RFA measurement during implant treatment is shown in Table 3.1.

Fourteen patients (eight males and six females, aged from 25 to 55 years) were involved, and they were scanned using CBCT (Promax 3-D, Planmeca, Finland). To maintain the same parameters for every patient on each stage of monitoring, the parameter scanning that consists of voltage, current, resolution, field of view (FOV), and patient's position is kept the same. Hence, those external factors were assumed not to affect the bone density measurements obtained from the CBCT images.

The position of patients during CBCT scanning was kept in a standing position with the head upright and positioned so that the intersection lines were straight horizontally and vertically through the centre of the region of interest. For all patients, CBCT images were taken with the following parameters: 84 kVp, 8 mA, 320-μm voxel resolution, and FOV 16 cm. All the recorded images are saved in the machine, and they will be transferred into MIMICS software for density estimation and further processing steps.

3.6.1 Validation of CBCT Scanning on Phantom

Density estimation from CBCT is one issue that wants to be investigated through this research. Hence the validation of CBCT is necessary that we need to do first. The CBCT validation was performed by using a known-density object. In this research, an object (ATOM Max Dental & Diagnostic Head Phantom, Model 711–HN, CIRS, Nederland) has been used. The density of this phantom is close to realistic human tissues and geometry. The energy of the CBCT was set to 50 keV–25 MeV. The model and scanning position of this model using CBCT scanning are shown in Fig. 3.11.

The material of Model 711-HN provides the physical of the male human head, and the size and structure represent the general condition of a male human head. The component of this model includes: the anthropomorphic anatomy including brain, bone, larynx, trachea, sinus, nasal cavities, and teeth. In this model, the bone is provided in both cortical and trabecular models. The structure of the dental system

Table 3.1 Evaluation measurement schedule during implant placement

Scanning	Pre-scanning	Stage 1	Stage 2	Stage 2
RFA scanning CBCT scanning	1 week before Implant (without RFA)	Immediate after implant	3 months after implant	4 months after implant (1 month post-crown)

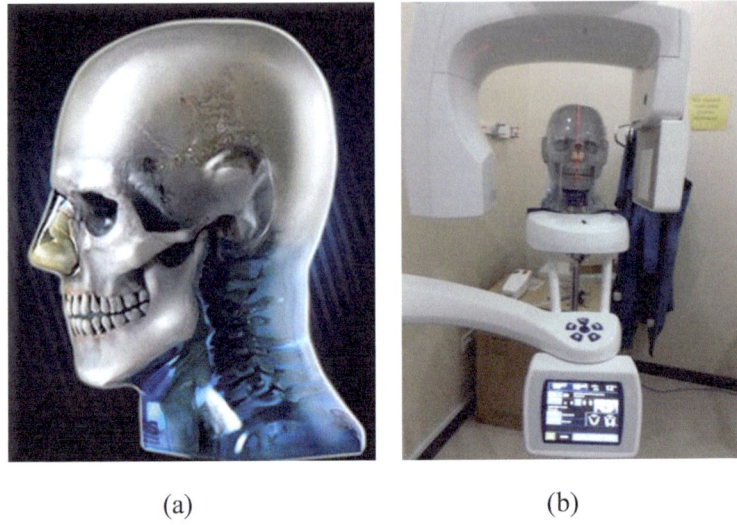

Fig. 3.11 a Model 711-HN and b the position on the CBCT scanning

is represented in detail including dentine, enamel, and root structure including the nerve. The densities of each element of this model are provided by the vendor, and it was provided in g/cc. The detail of the density of each element in this model is provided in Table 3.2.

Table 3.2 Density and material of elements of CIRS Model 711-HN

Description	Percentage by weight								Density, g/cc
	C	O	H	N	Ca	Mg	Cl	Others	
Soft tissue (blue)	64.36	20.57	9.07	6.00	0	0	0	0	1.103
Cortical bone	25.37	35.28	3.3	0.91	22.91	3.36	0.03	P (8.82)	1.91
Trabecular bone	54.84	25.28	7.46	1.3	8.84	2.12	0.11	0	1.197
Spinal disc	45.76	31.06	6.71	1.88	0	14.36	0.21	0	1.13
Spinal cord	56.21	24.78	8.46	1.53	0.25	8.41	0.18	P (0.12)	1.061
Brain tissue (average)	53.6	26.49	8.16	1.53	0	9.98	0.19	0	1.07
Tooth enamel	21.81	34.02	2.77	0.82	26.6	0	0.03	P (12.33)	2.04
Tooth dentin	35.35	29.41	4.51	1.23	19.84	0	0.04	S (0.31), Ba (1.31)	1.66
Sinus cavities (lung inhale)	65.89	19.26	8.59	3.52	1.01	0	1.69	P (12.33), S (0.31), Ba (0.33)	0.205

3.6 Dental Imaging and Implant Stability Measurement

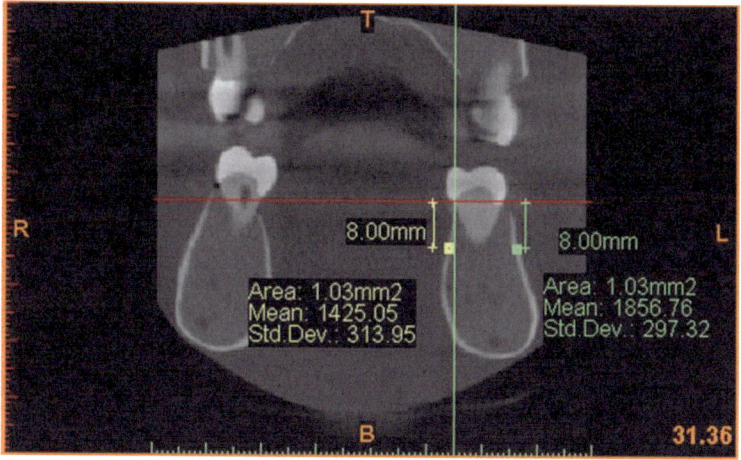

Fig. 3.12 CBCT scanning result of phantom 711-HN model and its density measurement on MIMICS software at 8 mm level (apically) from CEJ

Phantom is scanned five times with slice size 250 × 250 with increment of 0.320 mm and pixel size is 0.320 mm, field of view (FOV) is 8 cm, gantry tilt is 0.000, and number of slices is 250. The energy of CBCT was set to 90 KV, 122.99 mAs. The examples of CBCT scanning results at the target area are presented in Fig. 3.12.

For each CBCT scanning, the densities are measured in the same location. The obtained HU value from CBCT is plotted against actual density to get the relation between HU and density in gr/cc. This relationship is useful to estimate the density in gr/cc from CBCT which is measured in HU.

3.6.2 Validation of CBCT Scanning Using CT Scanning

The application of CT scanning for dental imaging has been widely practised in clinics; the grey value of CT image is associated with the density of the object. Objects with high density will appear as a high intensity, and inversely the object with low density will be identified as low intensity or even dark. The space of infection can be identified as a dark area. Because CT is an established technology for dental imaging, it is important to validate CBCT scanning with CT in terms of accuracy in determining density.

For this purpose, a model phantom (CIRS Model 711-HN) that was used in the CBCT scanning was scanned also with a CT machine. In comparison with CBCT scanning, CT scanning is varied with different angles as same as CBCT scanning. The phantom was scanned at 0°, 15°, and 30°. The positions of the phantom during scanning with different angles are shown in Fig. 3.13.

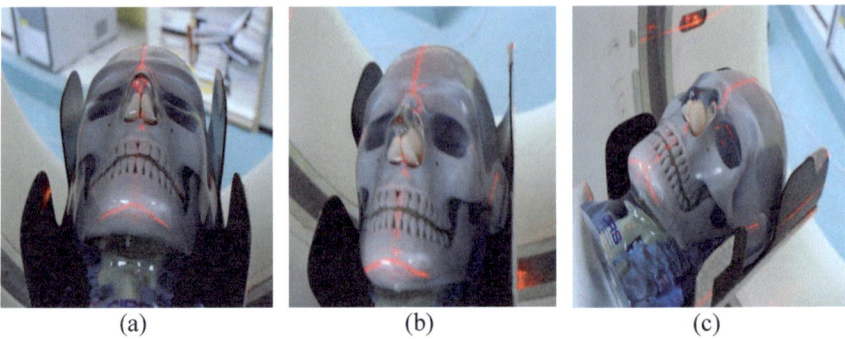

Fig. 3.13 CT scanning of phantom with different angles: **a** 0°, **b** 15°, and **c** 30°

3.6.3 *Measurement of Bone Density Using MIMICS Software*

The CBCT data that were stored in the database of the CBCT machine were retrieved into DICOM format and loaded into MIMICS software for bone evaluation purposes and further processing. During loading into MIMICS software, the coordinate system of image scanning is checked to make sure that the loading process does not change the coordinate system during the transformation.

The density measurements are conducted on MIMICS software by using density measurement tools. Bone density is estimated in two different point locations: buccal and palatal sides. Each density measurement location is evaluated at a different level measured from the cementoenamel junction (CEJ). For monitoring during treatment, density for each patient was measured at an 8 mm level from CEJ at the buccal and palatal side and it was measured for every stage of monitoring. Meanwhile, for repeatability study purposes, the density was evaluated at 6, 8, and 12 mm measured from CEJ.

3.6.4 *Measurement Implant Stability Using Resonance Frequency Analysis*

Implant stability which mechanically represents how well that implant tightened into the dental system was measured using a Resonance Frequency Analysis (RFA) instrument. The RFA device used is (Osstell, Integration Diagnostic AB, Göteborg, Sweden) with the SmartPeg abutment (Integration Diagnostic AB). Measurement of the implant stability using this instrument follows the protocol of Barewal et al. (2003) and Bischof et al. (2004) In this procedure, the SmartPegs were mounted on the implants and tightened by hand with a screw. The transducer was directly connected perpendicular to the implant as recommended by the manufacturer. The illustration of implant measurement using an RFA instrument is shown in Fig. 3.14.

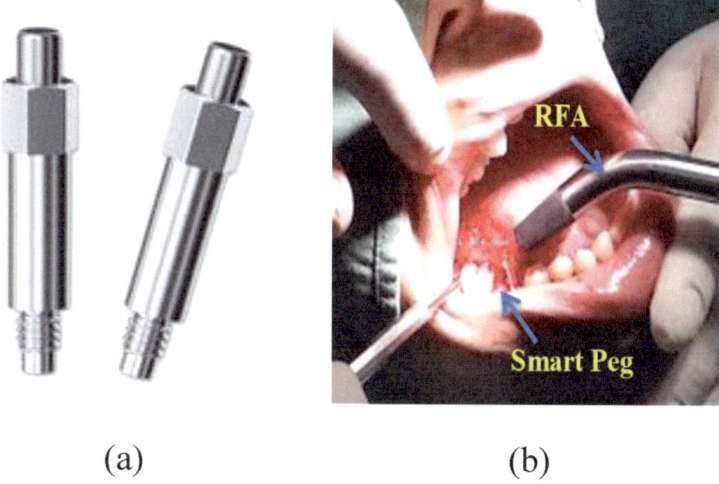

Fig. 3.14 **a** Implant with a SmartPeg to measure stability implant using RFA and **b** measurement of RFA from the buccal and lingual side

Measurements were taken twice in each direction: in the buccolingual direction from the buccal side and the lingual direction from the mesial side (Guler et al. 2013). RFA values were represented by a quantitative unit called the implant stability quotient (ISQ) on a scale from 1 to 100. The results were expressed in ISQ and averaged from all measurements for each implant.

Implant stability was measured three times for each patient, and the RFA measurement during implant treatment was taken following the schedule mentioned in Table 3.1. The RFA measurements are performed by researchers under the supervision of a Hospital USM specialist.

This measurement principle represents the loading process but on a very small scale of the force. A fixed lateral force is applied to the implant by a transducer, and its displacement is measured in terms of resonance frequency which is recorded back by a transducer. This method has proven useful for dental implants in research and clinical applications (Katsavrias 2009).

3.7 Finite Element Analysis (FEA) Study Preparation

A Finite Element Analysis (FEA) study is conducted to investigate the effect of loading numerically on dental implants and surrounding areas. Finite Element Analysis is performed by using ANSYS 14.0 software with the input being the object of the model prepared on MIMICS software and 3-matic software. The details of the scope of this FEA study will be explained in the next sub-chapter.

3.7.1 FEA Workflow

To perform Finite Element Analysis, some preparation steps are needed. Those steps are segmentation of the geometry of the object, meshing, material properties completion, simulation and analysis of the result. The workflow of steps is presented in Fig. 3.15.

In this research, the FEA was focused on a mandible jawbone-dental implant system that consists of an implant, crown, two neighbours' teeth and mandible bone which consists of cortical and trabecular. The bone of the mandible is assumed to consist of a cortical (outer part), and the rest is a trabecular (inner part). Segmentation of this bulk mandible model is performed on MIMICS software to construct all of each component. The illustration of the segmentation process is shown in Fig. 3.16.

MIMICS software version 17.0, manufactured by Materialize NV Technologielaan 15 BE-3001 Leuven, was used in segmentation. The segmentation steps are started by separating the hard tissue from the soft tissue using the threshold technique on HU of CBCT data. The cut-off HU for this separation is determined case by case; it was not the same from patient to patient. After hard tissue separated from soft tissue (Fig. 3.16a), the region of interest (ROI) is cropped to get only an area around the implant (Fig. 3.16b). On this bulk of ROI, every element that consists of two neighbour teeth, bone, implant, and crown was segmented using the Boolean operation technique (Fig. 3.16c–d). In the final stage, wrapping and smoothing are performed to get a smooth surface and closed volume for each of the objects. The

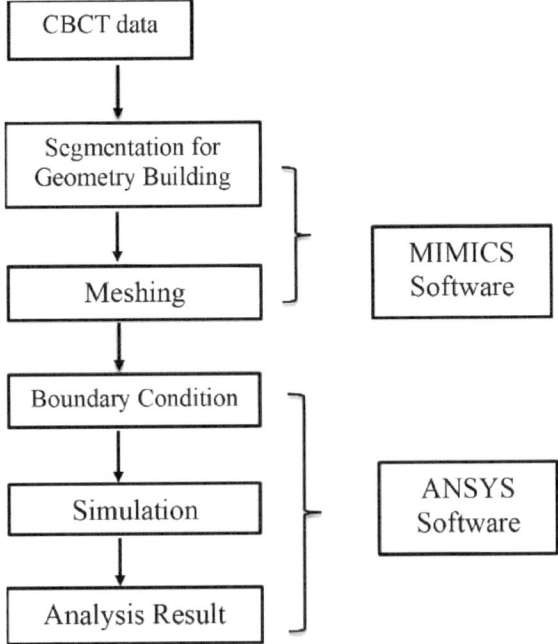

Fig. 3.15 Workflow of FEA study

3.7 Finite Element Analysis (FEA) Study Preparation

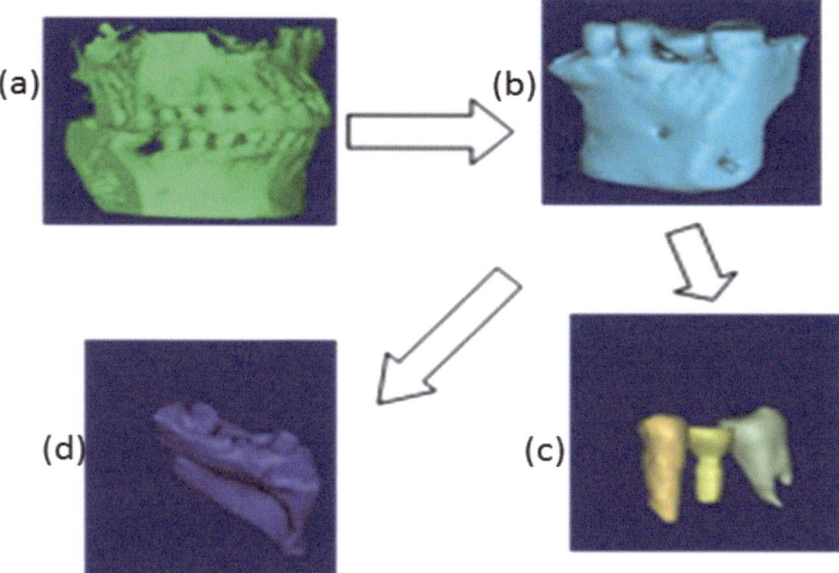

Fig. 3.16 Process of segmentation from CBCT data to construct 3-D objects

meshing of the object was performed using 3-Matic software, which is part of the MIMICS software

3.7.2 Material Assignment and Boundary Conditions

The next step after volume meshing is a material assignment step. All the elements in this model are regarded as homogeneous, isotropic, and linearly elastic materials (Hollister et al. 1992). The mechanical properties value required for material assignment on FEA study include density (ρ), Poisson's ratio (v), and Young's modulus (E) of each component and are adopted from published data. Those parameters are summarized in Table 3.3.

Table 3.3 Material properties for material assignment during FEA study

	Density (gr/cc)	Young modulus, E (GPa)	Poisson's ratio
Bone Cortical	2.17	13.7	0.3
Teeth	2.9	50	0.33
Implant (titanium)	4.51	105	0.37
Porcelain prosthesis	6.05	70	0.19

Boundary conditions in the interface between implant and bone were regarded as frictional with a certain friction coefficient adjusted. The rest, the contact between teeth and bone, and trabecular-cortical are regarded as bonded contact. The objective of this FEA study is to focus on the area surrounding the implant, including the area in the body implant, the bone surrounding the implant, and the neighbour tooth. The position and direction are referred to as a Cartesian system, X, Y, and Z coordinates. All the loading simulations which are represented by a given force are measured in the X, Y, and Z coordinates.

3.7.3 FEA Simulations

FEA simulations are conducted to investigate the relationship between loading processes in different conditions: pre- and post-crown conditions with stress distribution and micromotion of implant and neighbour tooth. The simulation consists of vertical loading, horizontal loading, and torque to represent the removal torque that might be generated during regular mastication. Types of loading are shown in Fig. 3.17.

In the vertical force simulation, a 200 N of force has been applied in the vertical direction. This simulation can be used to assess the stress distribution and micro motion that might be generated due to vertical force in the area surrounding the implant and neighbour teeth. The stress distribution and micro motion will be correlated with other properties of the implant site such as bone quality and quantity, the geometry of available space of the implant site and the height and width of the mandible.

In the horizontal force simulation, 200 N of horizontal force also was used same as vertical force simulation. However, the direction of force is different. In the horizontal simulation, force is directed in the horizontal direction from inside to outside of the mandible. This simulation is performed to estimate the healing capabilities within the bone-implant interface (Brunski et al. 2000). Even practically this test is destructive, and it is not practised in the clinical; however, studying numerically through FEA, the horizontal test also can show the ability and behaviour of the system in terms of stress distribution and micro motion because of loading process such as masticatory process and other unpredicted processes.

Removal torque simulation is conducted to investigate the torsional force necessary for unscrewing the implant (Johansson et al. 1998). In the clinic, the removal torque can be estimated and recorded using a torque manometer during implant insertion when the implant is tightened. The removal torque is a destructive test; hence it is not recommended to be applied practically in the clinic. However, the study numerically about this technique can give us information about how the resistance of the dental system to the shear force. In some references, the resistance of the implant on the removal torque also can be used as primary implant stability (O'Sullivan et al. 2000). The torque that was used for this simulation is about 200 Nm (Fig. 3.17c).

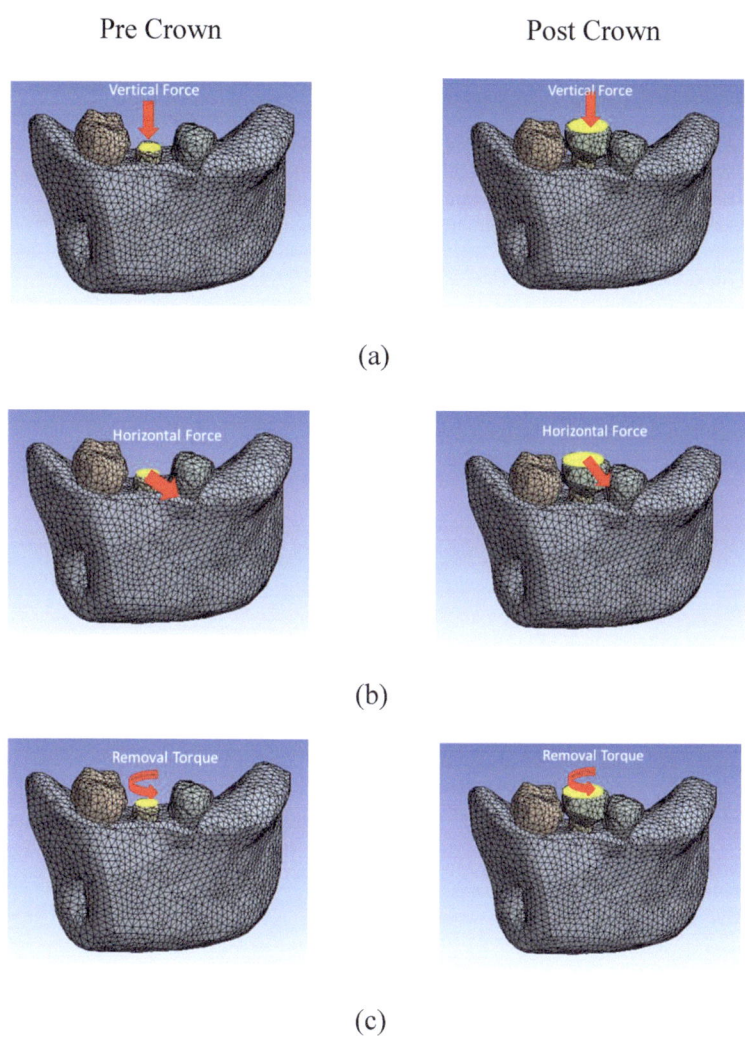

Fig. 3.17 Simulation of different loading: a vertical loading, a horizontal loading, and c. Removal torque, at pre-crown and post-crown conditions. Arrows show a force/torque location

References

- R.M. Barewal, T.W. Oates, N. Meredith, D.L. Cochran, Resonance frequency measurement of implant stability in vivo on implants with a sandblasted and acid-etched surface. Int. J. Oral Maxillofac. Implants **18**, 641–651 (2003)
- M. Bischof, R. Nedir, S. Szmukler-Moncler, J.P. Bernard, J. Samson, Implant stability measurement of delayed and immediately loaded implants during healing. Clin. Oral Implants Res. **15**, 529–539 (2004)

J.B. Brunski, D.A. Puleo, A. Nanci, Biomaterials and biomechanics of oral and maxillofacial implants: current status and future development. Int. J. Oral Maxillofac. Implants **15**, 15–26 (2000)

A.A. Guler, M. Sumer, I. Duran, E.O. Sandikci, N.T. Telcioglu, Resonance frequency analysis of 208 straumann dental implants during the healing period. J. Oral Implantol. **39**(2) (2013)

S.J. Hollister, J.M. Brennan, N. Kikuchi, *Computer Methods in Biomechanics and Biomedical Engineering* (Books and Journals Int. Ltd., Swansea, 1992), pp. 308–317

C.B. Johansson, C.H. Han, A. Wennerberg, T. Albrektsson, A quantitative comparison of machined commercially pure titanium and titanium-aluminum-vanadium implants in rabbit bone. Int. J. Oral Maxillofac. Implants **13**, 315–21 (1998)

G. Katsavrias, Reliability and validity of measuring implant stability with resonance frequency analysis, in *Master of Science in Dentistry* (Saint Louis University, Saint Louis, 2009)

M.J. O'Sullivan, M. Kyriakos, X. Zhu, M.R. Wick, P.E. Swanson, L.P. Dehner, P.A. Humphrey, J.D. Pfeifer, Malignant peripheral nerve sheath tumors with t(X;18). A pathologic and molecular genetic study. Mod. Pathol. **13**(11), 1253–1263 (2000)

Chapter 4
Biomechanical Assessment Based on Clinical Measurement

4.1 Accuracy and Repeatability Assessment

There is no doubt about the spatial accuracy of CBCT. During a clinical application, the spatial accuracy of CBCT is in the range of half a millimetre (Brullmann and Schulze 2015). However, the accuracy of CBCT in estimating the density is still questionable; hence, some experiment/testing to justify this accuracy is still needed. In the following, subchapter will be explained the experiment on the accuracy and repeatability of CBCT measurement in estimating the density.

Because of some reason, the position of the first and the next scanning cannot be maintained exactly. In this case, the different angle scanning might happen between two different scannings. Hence, it is important to know the effect of different angles on CBCT scanning, the response of the Hounsfield unit (HU) on different angles that might result during the repetition of scanning needs to be tested.

To evaluate the effect of different angles on HU values, a jaw phantom was scanned with three different angles: 0°, 15°, and 30°. As a reference, a CT scanning with similar angles is conducted on that phantom. This method allowed us to obtain comparable images of the same region of interest between CT and CBCT, and also it can understand the sensitivity of both techniques regarding the issue on different angle scanning.

To analyse the sensitivity of CBCT on different angle scanning, the protocol employed in this in vitro study consisted of an integrated sequence that involved several steps.

Statistical analysis to measure the difference from each angle scanning is performed by using SPSS software. The statistical description of each Hounsfield measurement consists of mean and standard deviation that are calculated using a t-test method. The confidence level was set to 95%. The quantitative data of each group are described with mean values and standard deviation (Table 4.1), and the differences in the density in HU are described in Table 4.2. (Group A1 vs. B1, Group A2 vs. B2, Group A3 vs. B3).

Table 4.1 Descriptive statistics: mean, standard deviation, and minimum and maximum of bone density values, defined as grey density values (VV)

Group	Mean	Maximum	Minimum	SD
A1	720.27	1160.78	298.89	276.40
A2	875.90	1525.22	487	312.62
A3	833.40	1293.91	557.11	230.94
B1	1086.29	1483.44	698.44	187.63
B2	1001.80	1309.78	609.67	175.50
B3	1071.19	1379.89	652.33	178.83

Table 4.2 Significance difference of each group

Group	Significant	Difference between means	Standard error
A1–A2	0.030*	155.63	45.81
A1–A3	0.206	113.13	45.39
B1–B2	0.091	84.49	85.02
B1–B3	0.532	15.09	73.23

*Statistically significant ($p \leq 0.05$)

4.2 Repeatability Measurement of Bone Density Using MIMICS Software

Bone density estimation which is based on HU of CBCT data is estimated by using MIMICS software. In this software, there are two methods available for density measurement: 2-D and 3-D tools. However, there is no literature review found that reported the repeatability of this measurement. Hence, before this method is applied for further analysis, the repeatability of the measurement of these methods needs to be tested. In this research, both available tools on MIMICS software are tested.

The error study has been conducted to test the repeatability of density measurement by involving 30 dental implant patients. On each CBCT of the patient, the 2-D and 3-D methods are used to estimate bone density. A 2-D method measurement is conducted with the steps: cementoenamel junction (CEJ) level determinations, defining the location of the target of measurement (8 mm from CEJ level), then creating the circle area and software calculating the density in the circle automatically. In the 3-D method, the steps were: threshold, segmentation of ROI, Boolean operation, and cropping mask. The density was calculated as the HU average of 3-D object obtained from the cropped mask. In every patient's data, only 3 locations are selected for measurement. These locations are selected either from: Incisor 1, Incisor 2, Canine, Premolar 1, Premolar 2, Molar 1, Molar 2, or Molar 3 of the mandible or maxilla. The point of measurement is located 8 mm from the CEJ level measured twice to reduce the error due to the inaccurate measurement. The results are an average of two measurements at the same location of measurement for each

4.2 Repeatability Measurement of Bone Density Using MIMICS Software

measurement; the HU is measured using the 2-D and 3-D methods. The total points of locations were eighty-eight (88) points, while the measurement results graphically are shown in Fig. 4.1.

The differences between measurement 1 and measurement 2 are analysed by using ANOVA in Microsoft Excel. To avoid bias in the result due to different methods (2-D and 3-D methods), the differences in measurement between 2-D and 3-D methods in every point measurement are calculated.

In both methods, the confidence levels were set to 95%. The statistical analysis result is shown in Table 4.3.

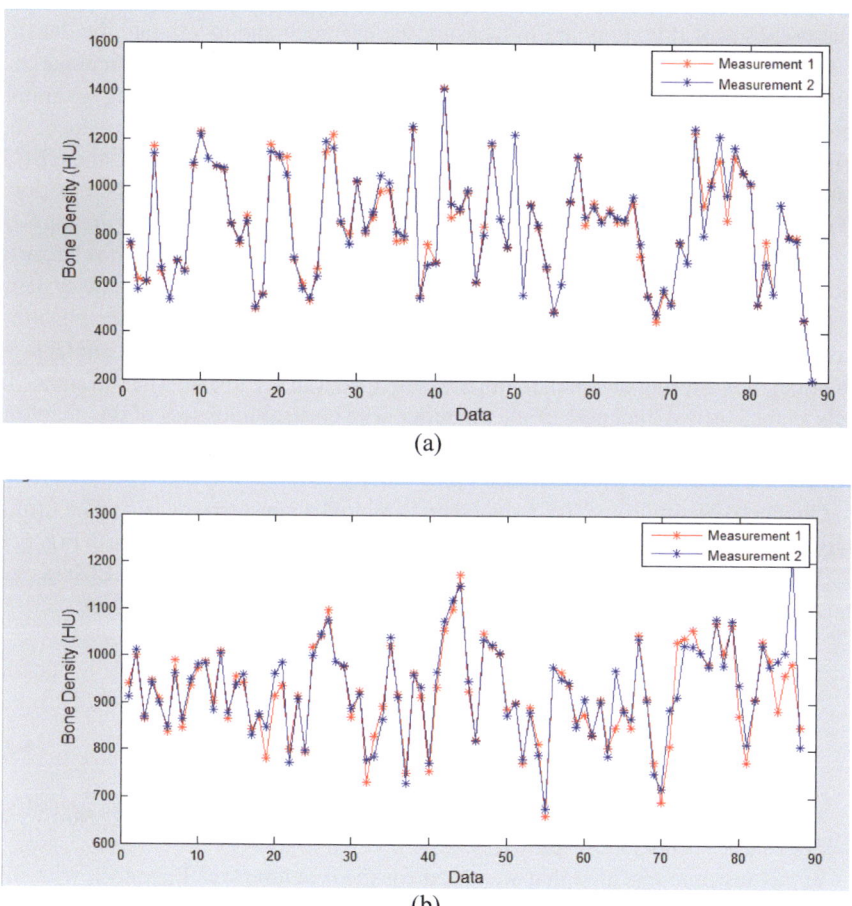

Fig. 4.1 Measured density in HU from CBCT data using **a** 2-D method and **b** 3-D method

Table 4.3 Repeatability and significant difference between measurement 1 and measurement 2 in 2-D and 3-D methods

Method	Repeatability (%)	p-value
2-D	99.05	8.66E−78
3-D	94.46	9.21E−38

4.3 Validation Density Measurement of CBCT Using Phantom

The purposes of this study are to measure the accuracy and to validate the density measurement from CBCT data. A jaw phantom with densities of anatomical parts is known to have been scanned five times by using a CBCT machine. During scanning, the position of the phantom is aligned in the patient scanning position; hence, the effect of different angle scanning is avoided, and it can be neglected. All the CBCT data then is loaded into MIMICS software for density measurement. The positions of measurements were located at the anatomical parts where the densities are known.

The quality of CBCT data is good and can represent the in vivo data; it is because the material of the phantom made as close to the density of the human jaw system. The main tissue such as the cortical, enamel, jawbone, and trabecular of the phantom has a density which is similar to the density of each human tissue. Weak artefacts as an effect of scattering due to hard objects still appear in the image. The difference is only in the cortical thickness of the phantom. The cortical thickness of the phantom is thinner compared with the human jaw. Examples of density measurement on MIMICS software are shown in Fig. 4.2.

Obtained density from CBCT data then is plotted against true density. The fitting curves were estimated with linear and logarithmic approximation as shown in Fig. 4.3.

The fitting curve equation for the relation between greyscale of CBCT as an independent variable (x) and density as the dependent variable (density) are:

$$\text{Density}(g/cc) = 0.0005x + 0.5171 \tag{4.1}$$

$$\text{Density}(g/cc) = 0.8564 \ln(x) - 4.8686. \tag{4.2}$$

The coefficient correlation (R) for the linear and logarithmic models are $R^2 = 0.7707$ and $R^2 = 0.9535$, respectively.

In this test, the machine that was used for the scanning was Planmeca with the energy of the X-ray was 84 kV 168.70 mAs. The result could be different from other researchers who used different machine specifications. Fang et al. (2006) reported that the relationship between density and HU obtained from the NewTom CBCT machine is linear. However, the location of the object during CBCT scanning using the NewTom CBCT machine does not affect the HU significantly (Lagrave et al., 2008).

4.4 Justification of Density Measurement Using CT Scanning

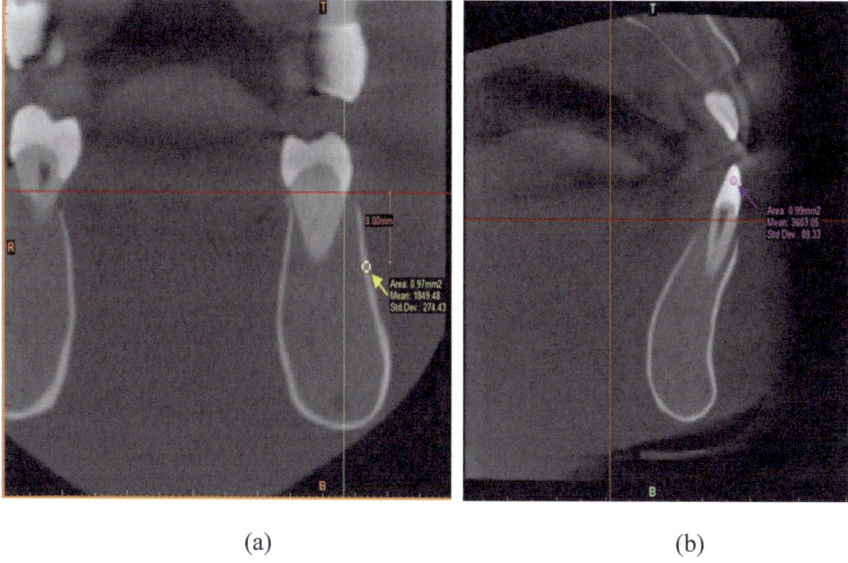

Fig. 4.2 Density measurement of phantom on CBCT data at: **a** cortical bone and **b** enamel

4.4 Justification of Density Measurement Using CT Scanning

To compare the CBCT reading with CT, the same object was scanned by using a CT machine. The data collection procedure and processing of CT are similar to CBCT explained in the earlier subchapter. The difference is only in the use of the source. The CT scanning used a GE Medical System CT machine with the energy of an X-ray set to 80 kV 30.94 mAs.

Density measurements on CT images are performed in the location as CBCT. There are eight measurement locations located in the anatomical part where the density is known. Measured density in HU of CT and the true density of different anatomical parts are presented in Table 4.4

The correlation between true density and HU of CT is approximated with linear and logarithmic approximation as shown in Fig. 4.4.

From this measurement, the density of the object can be correlated with the HU of the CT with the formula:

$$\text{Density (g/cc)} = 0.0005x + 0.5079 \tag{4.3}$$

Or:

$$\text{Density (g/cc)} = 0.8634 \ln(x) - 4.9815. \tag{4.4}$$

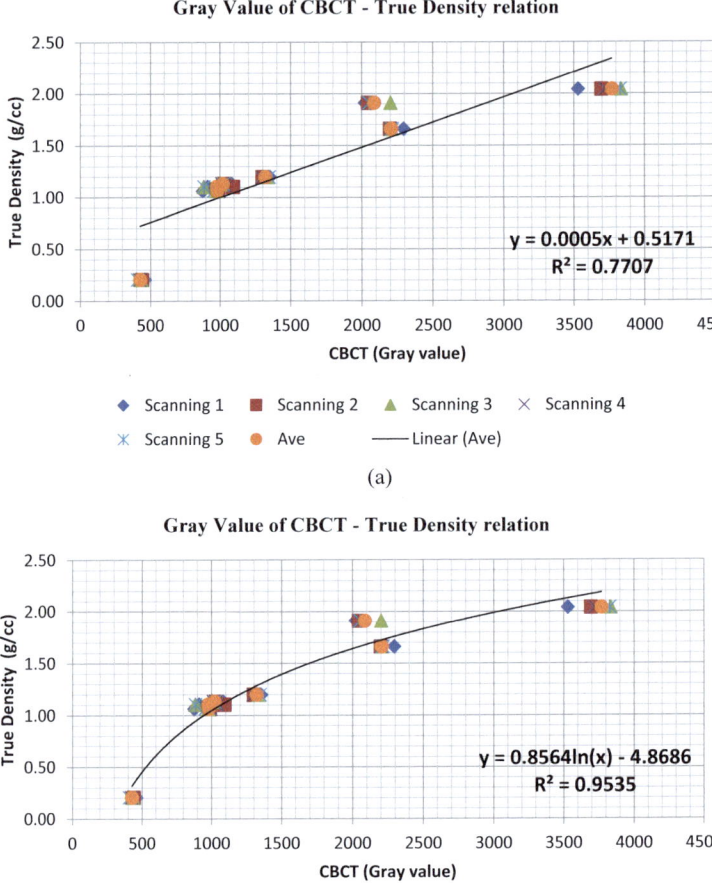

Fig. 4.3 Relation between true density and greyscale of CBCT of the phantom **a** linear regression, **b** logarithmic regression

A linear and logarithmic approximation in Eq. (4.3) has coefficient correlation $R^2 = 0.759$ and $R^2 = 0.9362$, respectively.

The direct correlation between greyscale of CBCT and CT can be obtained from the cross-plot as shown in Fig. 4.5.

Paired t-tests have been applied to these datasets to measure the significant differences among them. The mean difference between CBCT and CT reading is about 110.88 where the CT reading is higher than CBCT. The difference is not significant statistically ($p > 0.05$). The reading of CBCT is lower than CT; in this result, the

Table 4.4 Measured density based on CT data and true density of the object

No	Anatomy	Density (g/cc)	Density (HU)
1	Soft tissue	1.103	1052.51
2	Cortical bone	1.91	2098.09
3	Trabecular bone	1.197	1442
4	Tooth enamel	2.04	3950.29
5	Tooth dentin	1.66	2508.73
6	Sinus cavity	0.205	470.48
7	Spinal cord	1.061	1056.78
8	Spinal disc	1.13	1085.23

differences range from 33 to 189 HU. Statistically, the readings of CBCT and CT are different significantly ($p < 0.05$).

4.5 Discussion

In the same anatomy, the density of the phantom is homogenous. The different HU readings in different scanning will represent a reading error of HU directly. The results show that in the same location, the density of the phantom is detected differently on each different angle scanning either on CBCT or CT. In the 0°, 15°, and 30° of CT scanning, the HU is not too much different compared with CBCT. The difference in HU is not linear with the changes on angle scanning. The different angle scanning will give some difference on the positioning of certain part of the object measured from the source of X-ray. It shows that CBCT scanning is more sensitive to the position of the object compared with CT.

The difference on the sensitivity of CBCT and CT scanning regarding the different angle scanning can be understood from the acquisition configuration of both methods (Fig. 4.6a). The basic difference is that the CBCT is using a cone beam source; meanwhile, the CT is using a fan beam source. The different type of source will be responsible to the error resulted by misalignment of scanning. For example, if the object is located at point A in the first scanning. Then in the next scanning, that object underwent a rotation with a different angle θ without translation. That object from the first position at point A (in the first scanning) will be projected into a new position at A′ (in the second scanning). The difference distance from the object to the source between the two scans is **r** × **θ**. Not only the HU, which is captured in the detector, this difference distance also will affect the shape of resulted image.

Unlike CT scanning, the different angle scanning on CBCT is more complex because of the source type. In the CBCT, rotation on the object will produce an error in all directions, including the x and y positions. This different angle on CBCT at least triggers the errors in vertical shift, horizontal transversal shift from the central ray and horizontal longitudinal shift along the central ray direction. These three deviations

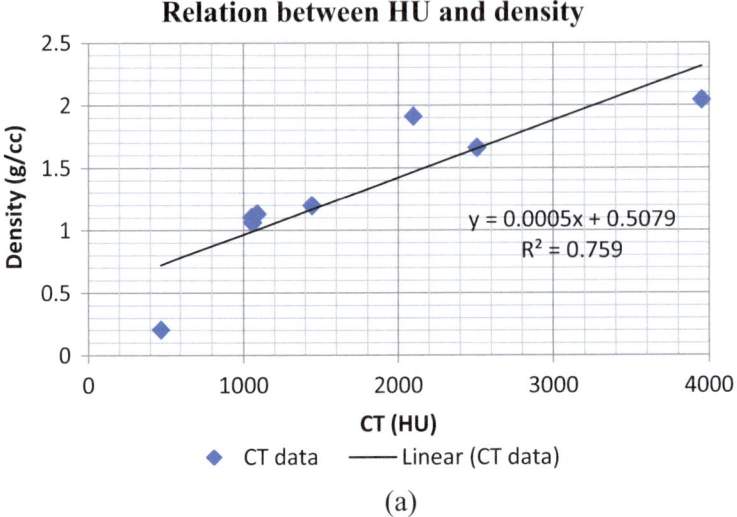

(a)

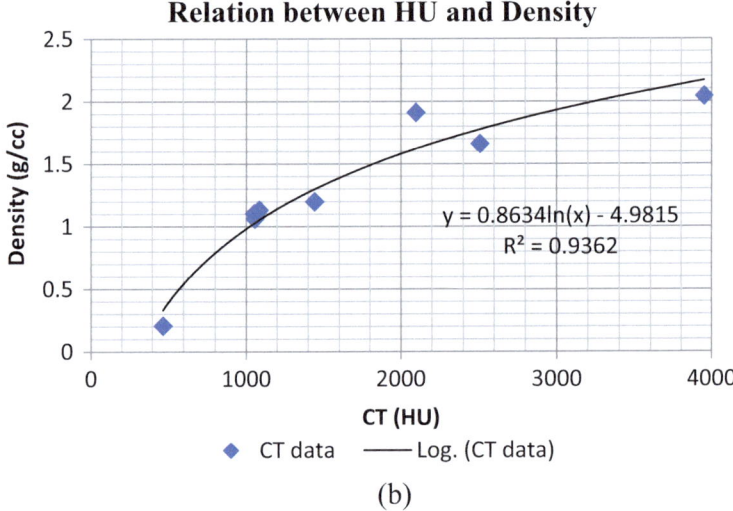

(b)

Fig. 4.4 Curve estimation of density—HU relation from CT scanning: **a** linear and **b** logarithmic

can produce not only the reading of HU on the plane detector projection but also cause an error in the reconstruction of shape. In this case, the X-ray of CBCT will be scattered stronger than CT; hence, the artefact on CBCT can be triggered more strongly than CT.

The error due to misalignment on CT and CBCT scanning has been tried to be compensated. Yi Sun [12] suggests that this error should be reduced before image

4.5 Discussion

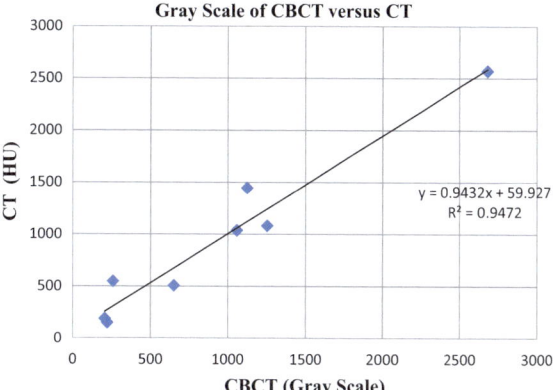

Fig. 4.5 Cross-plot between greyscale of CBCT and HU of CT

reconstruction. Not only the error in the value of HU, but the different angle scanning also affect the shape of image projection as reported by Ford et al. In our investigation, the result showed that HU in the CBCT and the CT resulted by using three different angles (0°, 15°, 30°) scanning produced the error in HU reading, but there are no comparable reports in the literature.

CBCT is a technique widely used in oral and maxillofacial surgery, especially for bone density and site of implant assessment. However, because of some limitations of the CBCT technique, the interpretation of this method needs to be done carefully. Some efforts to maintain the same position during CBCT scanning in the series of monitoring need to be highly considered. Hence the error due to misalignment/positioning during scanning can be minimized.

CBCT is a relevant radiographic diagnostic tool for monitoring dental implant treatment because this method can offer additional information on periodontal progression and disease which is difficult to obtain by other clinical methods such as probing depth measurements. However, some limitations on it including the quality of an instrument, instrument calibration, the technical aspect during data collection, and the limitations of the methods that are used also need to be considered to avoid miss diagnostics.

The repeatability and radiographic accuracy of CBCT have been tested by some authors. Gelaude et.al. (2016) tested the accuracy of CBCT in determining the distances between facial landmarks using MIMICS software. Their results showed that the 3-D surface accuracy of CBCT scans segmented with Mimics software is high. However, there is no literature on the repeatability measurements of density estimation using the same technique. As a comparison, Kim et al. (2017) used 3-D software to measure the repeatability of CBCT in defining the size of the mandibular arch. Their result demonstrated that the CBCT measurements are stable.

Uncertainty can arise when the exact area selection cannot be achieved the same. Some difficulties in defining the location of the samples as matching as previous will contribute to the error in repeatability results. It is because the CBCT has high lateral resolution; the different locations will associate with different anatomy, and

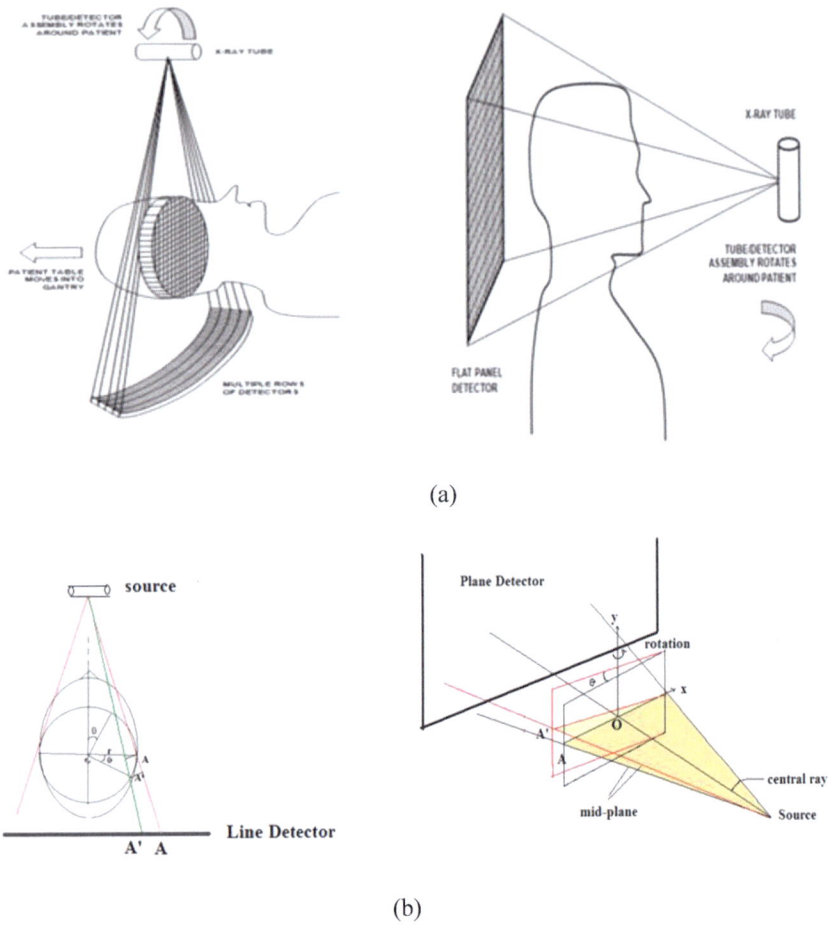

Fig. 4.6 Difference between CT and CBCT: **a** configuration of fan beam (left) and cone beam source (right) and **b** effect of misalignment on CT (left) and CBCT (right)

hence the density will be shown differently. In both methods, either 2-D or 3-D method, the density is calculated as an average value of HU of the selected area. The selection either on the 2-D or 3-D method is performed manually; hence, the error in the repeatability will be contributed by this factor. Another problem that might contribute to the error is like poor identification of anatomical landmarks such as CEJ as a reference in determining the level of measurement location.

The difference between measurement 1 and measurement 2 in both methods is not significant statistically ($p \geq 0.05$). However, the density measurement by utilizing the 2-D method produced a mean difference (mean different $= 1.55$ HU, $p = 0.66$) lesser than the 3-D method (mean different $= 5.35$, $p = 0.15$). Understandably, the error is due to the missing position in the 3-D method bigger than in the 2-D method.

4.5 Discussion

Hence, to avoid more errors due to repeatability measurement, a 2-D method is used for further analysis in this research.

The essential assessment of using CBCT data is to evaluate the quality of bone in terms of bone density which is predicted to have a good correlation with the success rate of dental implants. A CBCT scanning can provide the HU which can be related to the density of tissue. However, the true density in gr/cc of the object is questionable to be predicted from the CBCT data. To answer this uncertainty, the validation of CBCT scanning in determining the density has been conducted.

Before analysing the bone density changes during implant treatment, the accuracies of the instrumentation and measurements of density measurement needed to be validated. In this research, a phantom that consists of several kinds of materials: pure water, aluminium, dental composite resin, high-density acrylic, and dental utility wax with specific densities was used to validate the consistency of CBCT scanning in determining the density. The HU at a given location that the true density of that anatomical part is known is measured. The measurement is located at: soft tissue, cortical bone, trabecular bone, tooth enamel, tooth dentin, sinus cavity, spinal cord, and spinal disc.

The obtained results show that the density measured in greyscale from CBCT correlates with the density of the object. The true density in g/cc can be approximated by the HU of CBCT. The coefficient correlation (R) of a linear regression approximation is 0.77, while the logarithmic approximation is 0.95. This value shows that the estimation of density from HU of CBCT is reliable enough. However, the imprecision between HU of CBCT with true density may be attributed to some factors such as the characteristic of CBCT that can generate the scattering on the hard tissue especially when the cone angle is widened, miss-position of the measurement and artefact due to a combination of scattered radiation and beam hardening. Even the validation of phantom studies is questionable, but because this phantom can replicate the human jaw; hence, clinical conditions such as generated artefacts also can be accommodated. This study can give the evidence that HU obtained from CBCT is linear with density; however, the correlation coefficient is still low (0.77). Better density estimation from HU of CBCT can be optimized by using logarithmic relation which gives a higher correlation coefficient (0.95). By using this equation as formulated in Eq. (4.2), the HU of CBCT can be transformed into true density in the g/cc unit. More sample measurement/validation is still needed to improve the correlation between HU readings with the true density.

Pauwels et al. (2013) investigated the variability of the grey value of CBCT for density estimation and compared it with the multislice CT (MSCT). Their results showed that CBCT and MSCT have correlation coefficients which are generally high; most CBCT protocols showed an excellent linear fit between the CBCT and MSCT grey values. However, they suggested that the density prediction cannot be solely based on MSCT.

In this research, a comparison between CT and CBCT scanning has been conducted to investigate the reliability of CBCT as a tool for bone density estimation. By using the same object with the same protocol, the comparison between CT and CBCT scanning is possible to be done. Same as CBCT, the CT scanning also gives a good correlation between HU with the true density. The linear and logarithmic approximation has significant difference; the coefficient correlation of linear regression is 0.78 and logarithmic regression is 0.94. It shows us that the logarithmic approximation is better for density estimation from HU of CT. This means that the true density of bone should be calculated using a logarithmic formulation rather than linear regression.

Compared with CBCT, the CT reading is a bit higher than CBCT with a mean difference of about 110.88 HU with a standard error mean of 32.94. The difference between CBCT reading with CT is statistically significant ($p = 0.012$ at confidence level $= 95\%$). The result showed that the use of CBCT for density prediction on the jawbone is as acceptable as the CT method, but the reading from CBCT bit lower than CT.

References

D. Brullmann, W.R.K. Schulze, Spatial resolution in CBCT machines for dental/maxillofacial applications—what do we know today? Dentomaxillofac. Radiol. **46**, 20140204 (2015)

Y. Fang, J. Carey, R.W. Toogood, G.V. Packota, P.W. Major, Density conversion factor determined using a cone-beam computed tomography unit NewTom QR-DVT 9000. Dentomaxillofac. Radiol. 407–409 (2006)

F. Gelaude, A.Z.S. Brugge-oostende, Accuracy and repeatability of cone-beam computed tomography (CBCT) measurements used in the determination of facial indices in the laboratory setup. J. Cranio-Maxillofac. Surg. **37**(1), 18–23 (2016)

S.R. Kim, C.M. Kim, I.D. Jeong, W.C. Kim, H.Y. Kim, J.H. Kim, Evaluation of accuracy and repeatability using CBCT and a dental scanner by means of 3D software. Int J. Comput. Dent. **20**(1), 65–73 (2017). PMID: 28294206

M.O. Lagrave, J. Care, M. Ben-Zvi, G. Packota, P. Major, Effect of object location on the density measurement and Hounsfield conversion in a NewTom 3G cone beam computed tomography unit. Dentomaxillofac. Radiol. **2007**, 305–308 (2008)

R. Pauwels, O. Nackaerts, N. Bellaiche, H. Stamatakis, K. Tsiklakis, Variability of dental cone beam CT grey values for density estimations. Br. J. Radiol. **86** (2013)

Chapter 5
Application of CBCT Data for Bone Quality and Implant Stability Monitoring and Correlation

5.1 Application of CBCT Data for Bone Quality and Quantity Assessment

One of the factors that determine the success of dental implant treatment is bone quality and quantity of patients. Therefore, evaluation of the bone quantity and quality as important prerequisites of predictable implant success (Song et al. 2009). The main objective of this study is to evaluate the bone quality (density) and quantity (volume) of dental implant patients during implant treatment based on CBCT data.

To understand the behaviour of bone quality and quantity during implant treatment, this study involved ten dental implant patients consisting of 5 males and 5 females with a range of age from 25 to 55 years old. All the patients had in sufficient level of oral hygiene; there is no infection was reported during the period of treatment, and each patient was treated with the (MegaGen Implant Co., Ltd., South Korea) for replacing one of the edentulous molar teeth on the mandible. The dental implant treatments were performed by specialists in USM hospital.

5.2 Bone Density Evaluation of Pre- and Post-Crown

Evaluations of density during dental implant treatment are performed in three stages. Stage 1 is performed by scanning the patient on the same day after implant placement. This condition is referred to as the initial condition where the primary implant is measured and bone mechanism as a response of implant insertion is still not started yet. Stage 2 is performed about 3 months after implant placement. In this stage, the crown has been installed and measurements are performed on the same day with the crown installation. The scanning is performed to measure the secondary implant stability and investigate the response of the jaw on implant insertion in terms of density changes. Here, it is assumed that the bone mechanism has been started.

Stage 3 is taken 4 months after implant placement or 1 month after crown installation; this stage is conducted to measure the implant stability after loading and bone density changes as a progress after the implant underwent a loading due to crown installation and external loading due to regular mastication.

In every stage, bone density is estimated based on CBCT data. The CBCT machine (Promax 3-D, Planmeca, Finland) was used to monitor the density of the jawbone. For all patients, the CBCT machine was set with the voltage, current, resolution, and field of view (FOV) as the same in every stage measurement. The CBCT parameters are 84 kVp, 8 mA, 320-μm voxel resolution, and FOV of 16 cm.

To minimize the effects of different angles on the resulting image, the position of the patient is maintained the same in every scanning by placing the patient in the standing position with the head upright. The intersection lines from a light control were set straight horizontally and vertically through the centre of the region of interest Fig. 5.1.

The density measurements are performed in MIMICS software by using 2-D method. The locations of density measurement are selected near to implant at level 8 mm from the cementoenamel junction (CEJ) level. To avoid the reading error due to repetition of scanning, one location that is far from the implant site is measured. The change of density in this location is used for reading correction of measurement in the site implant. Every measurement is performed in the buccal and palatal site, and the results are averaged.

The density changes during monitoring stages in terms of greyscale are shown in Fig. 5.2a. The estimated density in gr/cc is calculated by using Eq. (4.2) (Density (g/cc) = $0.8564\ln(x) - 4.8686$) and is presented in Fig. 5.2b. Most patients undergo a loss of density from stage 1 to stage 2 and increase again from stage 2 to stage 3.

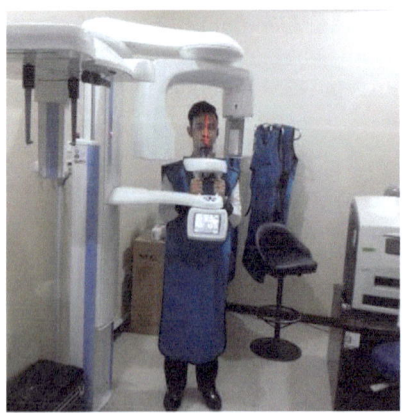 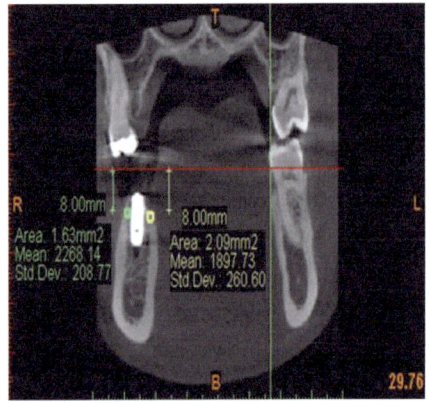

(a) (b)

Fig. 5.1 **a** Standing position of the patient during CBCT scanning and **b** density estimation from CBCT data in Mimics software

5.3 Cortical Thickness and Available Space Measurement

Fig. 5.2 Density of site implant during monitoring stage. **a** In greyscale and **b** In gr/cc

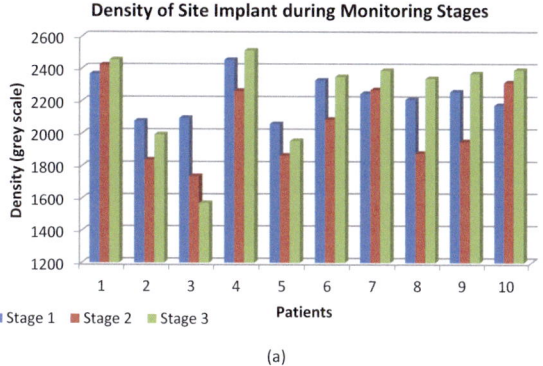

(a)

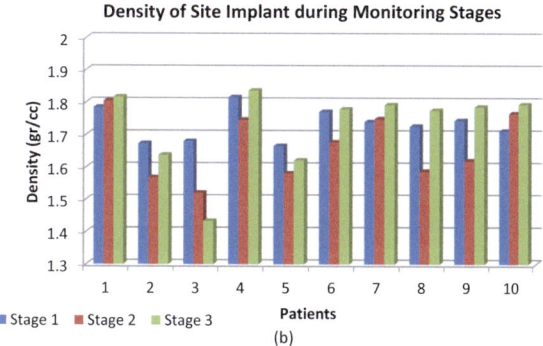

(b)

Only 3 patients have undergone the density increase since beginning, and 1 patient has density decreasing.

The statistical analysis including correlation and comparisons was performed based on paired t-tests using the SPSS 20 software package. The statistical significance has been determined by setting the confidence levels to 95%.

In general, the mean density decreased from stage 1 to stage 2 (significant $P = 0.17$) and increased again from stage 2 to stage 3 (significant $P = 0.122$) as shown in Fig. 3 (red-bold line). Statistically, the bone density decreased significantly from stage 1 to stage 2 and increased not significantly from stage 2 to stage 3.

5.3 Cortical Thickness and Available Space Measurement

The cortical thickness measurement is performed on CBCT data by using a tool that is available on MIMICS software. The location of cortical thickness evaluation is the same as bone density evaluation and near to implant at level 8 mm from cementoenamel junction (CEJ) level. The height and width of the mandibular were

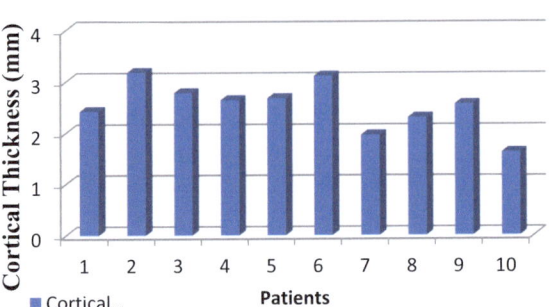

Fig. 5.3 Measured cortical thickness from CBCT data

measured in the location close to the implant. The cortical thickness is measured only one time for each patient; that is only in stage 1.

The cortical measurements are performed in two different locations, the buccal and lingual sides. Hence, the representative cortical thickness is an average of those measurements. The cortical thickness of all patients is shown in Fig. 5.3.

The width and height of available space for the implant site are measured around the implant. The width of the jaw is measured from the buccal to the lingual sides at an 8 mm level from CEJ with the line measurement perpendicular to the jaw. While the height of the jaw is measured from top to bottom of jaw vertically, the results are shown in Fig. 5.4.

Available space for the implant site is evaluated by calculating the volume of the jaw around the implant site. The volume is calculated by multiplying the height by the width and cortical thickness of the jaw. The results are shown in Fig. 5.5.

The descriptive statistics on those characteristic available spaces for the implant are analysed by using SPSS software. The results are shown Table 5.1.

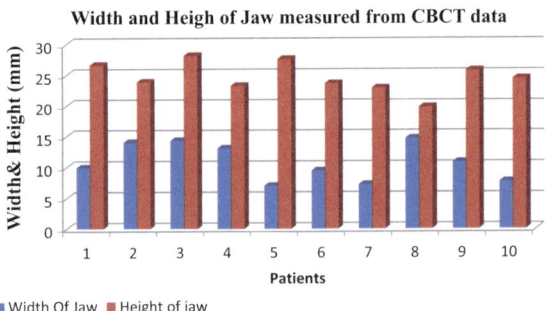

Fig. 5.4 Width and height of jaw measured from CBCT data

5.4 Discussions

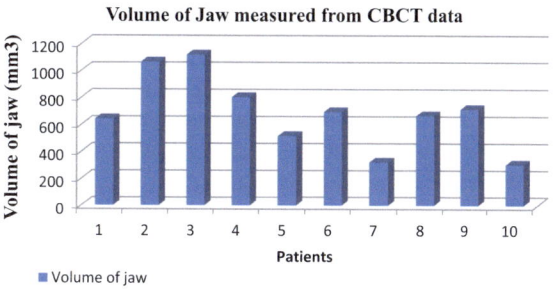

Fig. 5.5 Volume of jaw around implant site, measured from CBCT data

Table 5.1 Descriptive statistics of available space around the site implant

Descriptive statistics

	N	Range	Minimum	Maximum	Mean	Std. deviation
Cortical (mm)	10	1.56	1.63	3.18	2.5255	0.47878
Width (mm)	10	7.66	7.01	14.67	10.8320	3.02487
Height (mm)	10	8.36	19.80	28.16	24.6280	2.50068
Volume (mm^3)	10	815.11	305.93	1121.04	686.6552	269.84733

5.4 Discussions

Comprehensive evaluations on intra-oral (dimension of available space for implant placement) must be done, not only to evaluate whether it is possible to place the implant, but also to define the suitable implant size. In this research, evaluations of available space for implant have been done based on radiographic examination using CBCT data. Evaluations include assessment of cortical thickness, width, height, and volume of jawbone for every patient. The purpose of this evaluation is to know the correlation between quantities of bone with implant stability which will be explained in the next subchapter.

The cortical thicknesses have been evaluated for buccal and lingual. In general, the cortical thickness in the buccal location is higher than lingual. The mean cortical thickness in buccal and lingual are 2.81 mm and 2.24 mm respectively. Hence, based on the statistical calculation the difference is significant statistically ($p = 0.164$). The minimum cortical thickness for implant placement is about 1 mm (Shenoy 2012); in general, the minimum requirement for cortical thickness is achieved in every patient. There are only two patients who have cortical thicknesses less than 2 mm but still more than 1 mm.

In the mandible, the minimum height of available jaw to accommodate the shortest implants (7 mm) is 9 mm, at least 2 mm of additional bone is required to maintain a safe distance from the inferior alveolar canal (Goodacre and Anderson 2016). However, the shorter length of the implant will decrease the success of the implant. In this research, the implants that are used are 11 mm in length except for one patient who used an implant with a length of 10 mm. Hence, the minimum available height of the jaw around the implant site is about 12–13 mm. The results show that the height average of the jaw is about 24.63 mm with a minimum height is 19.80 mm and a maximum is 28.16 mm. All patients have available height of jaw that is more than the minimum requirement. The available space of the mandible for the implant site is greater than the maxilla (Mittal, Jindal and Sandeep 2016) (Genisa et al., 2017).

Adequate bone width is evaluated to know the size related to the diameter of the implant. The minimum requirements of bone width are dependent on the diameter of the implant. For example, an implant with a diameter of 4 mm needs a width of bone at least 6 mm. The minimum available bone width between implant to bone is 1 mm on either side (Shenoy 2012). In this research, the diameter of all used implants is 4 mm except for one patient-used implant with 4.5 mm. The results show that the average bone width is 10.83 mm with a minimum is 7.01 mm and a maximum is 14.67 mm. In terms of adequate bone width, all patients have bone width more than the minimum requirement.

5.5 Implant Stability Monitoring and Correlation

The successes of osseointegration in dental implant treatment are highly determined by implant stability. Monitoring continuously on implant stability is required to understand the progress on how bone will respond to the implant insertion. Clinically implant stability is divided into two types: primary implant stability which is measured on the same day after implant surgery and secondary implant stability which is measured after a certain time when the regeneration/remodelling of bone has started.

The purpose of monitoring implant stability during implant treatment in this research is to understand the behaviour of bone to respond to the loading due to internal and external loading and correlate it with other main parameters such as bone quality and quantity. Perhaps, investigating the behaviour of implant stability during implant treatment and correlating it with the quality and quantity of bone during treatment can help us to understand the bone mechanism/remodelling.

5.5.1 Implant Stability Measurement

Implant stability during implant treatment is measured by using Osstell^{tm} mentor device (Integration Diagnostic AB, Goteborg, Sweden) with the SmartPeg abutment from the same manufacturer. This instrument used the Resonance Frequency Analysis (RFA) concept and provided the implant stability of the object as an Implant Stability Quotient (ISQ) with a scale ranging from 0 to 100 (Brouwers et al. 2009). Monitoring stages of implant stability follow the stages of density monitoring, which is divided into three stages: stage 1, stage 2, and stage 3 which are explained in the previous subchapter.

5.5 Implant Stability Monitoring and Correlation

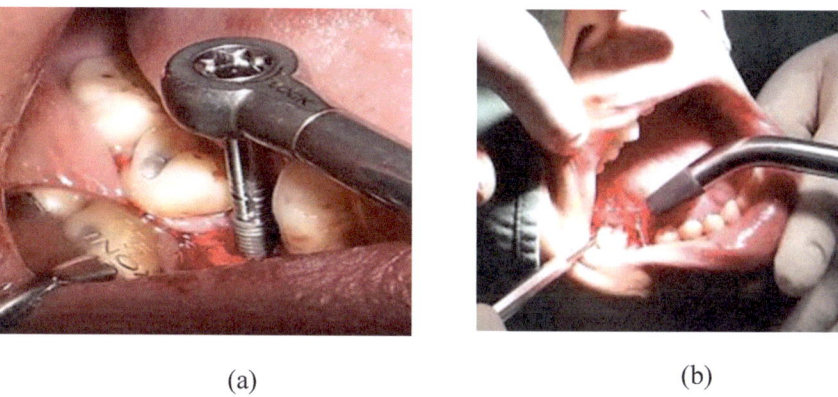

(a)   (b)

Fig. 5.6 **a** Implant insertion surgery, **b** implant stability measurement using RFA Osstell mentor device

The protocols that are used during measurement follow the Barewal (3) and Bischof (6) protocol (Ersanli et al. 2005). To avoid errors due to less coupling between instruments with the object, the SmartPeg was mounted into the implant, and it was tightened. The transducer with L-shaped is connected directly perpendicular to the implant following the procedure recommended by the manufacturer Fig. 5.6.

All measurements are performed two times by placing the probe on the buccal and palatal sides. Average implant stability from buccal and palatal measurements is used to represent the total implant stability of the patient in every stage.

Statistical analysis of difference measurement between stages is performed by using SPSS. The confidence level was set at 95%. The results show that there was some increment in implant stability from stage 1 to stage 2 to stage 3. The average implant stability has increased from 68.85 ISQ to 77.80 ISQ in stage 1 to stage 2 with a significant difference of $p = 0.033$ and from stage 2 to stage 3, the average implant stability increased from 77.80 ISQ to 82.17 ISQ with a significant difference of $p = 0.016$.

5.5.2 Correlation Between Bone Quality/Quantity and Implant Stability

Implant stability as a main factor for the success of implant integration and density has been monitored in three stages. To know the main factor that affects implant stability, some variables (cortical thickness, bone width and height, and estimated density) are correlated to implant stability. Because two types of implant stabilities are measured during monitoring, the correlation is conducted differently. Density in the initial condition (stage 1), cortical thickness, bone width, and height are correlated with primary implant stability. Meanwhile, the secondary implant stability is correlated

Fig. 5.7 Plot between density and primary implant stability that are measured in stage 1

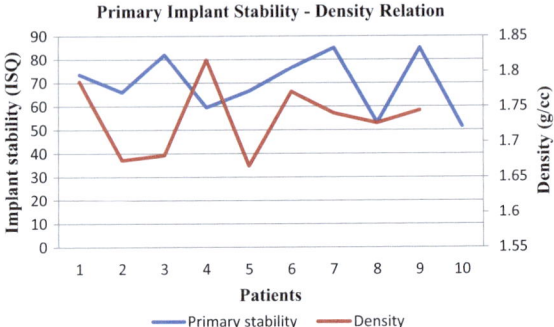

only with respective densities that are measured in each stage. It is assumed that the primary implant stability is associated with the mechanical engagement of an implant with the surrounding bone, while the secondary implant stability is assumed related to bone regeneration and remodelling process after a certain period. The mechanism of the jawbone in responding to the implant insertion which is measured as primary and secondary implant stability is not simple. Many factors can affect the primary and secondary implant stability such as the situation of surrounding tissues, bone quantity and quality, implant geometry, and surgical technique. The secondary implant stability is affected by primary implant stability itself, bone remodelling and implant surface (Javed et al. 2013).

Correlations between variables are tested in SPSS software by using Pearson's correlation one-tailed test. The relation between density estimated from CBCT data and primary implant stability is shown in Fig. 5.7. Statistical analysis showed that the correlation of each variable has a low Pearson's correlation coefficient ($r = 0.031$) with significant $p = 0.466$. This result shows that there is no significant correlation between estimated densities from CBCT with primary implant stability obtained from RFA measurement.

Primary implant stability correlations with other parameters of available space of the jaw are shown in Fig. 5.8. The figure showed that the primary implant stability is not following the trend of densities (Fig. 5.9a). However, bone height and cortical thickness patterns are like the implant stability variation (Fig. 5.9b, c). Implant stabilities are inversely proportional to angle insertion; patients with high implant stability have higher implant insertion angle. In other words, if an implant is placed with oblique, the implant stability will be lesser (Fig. 5.9d).

Secondary implant stability is measured in stage 2 and stage 3. The whole implant stability and density obtained from CBCT during treatment are presented in Fig. 5.9. SPSS analysis on correlation based on Pearson's correlation shows that the correlation between density and secondary implant stability at stage 2 is not significant statistically (Pearson's correlation $= 0.205$, $p = 0.285$) and in stage 3 not significant statistically (Pearson's correlation $= -0.018$, $p = 0.482$).

5.5 Implant Stability Monitoring and Correlation

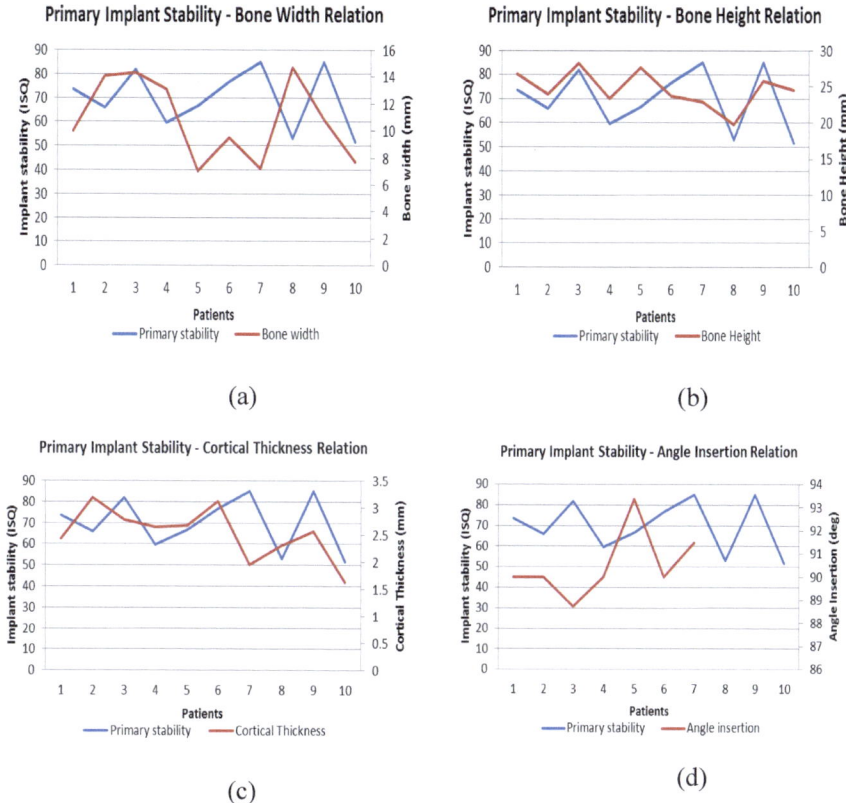

Fig. 5.8 Primary implant stability cross-plot against **a** bone width, **b** bone height, **c** cortical thickness, and **d** angle insertion

5.5.3 Discussions

Insertion of a dental implant into the jaw system will change the biomechanical dental system. Integration of biological processes of dental system and mechanical behaviour needs to be considered during the including the environment of surrounding implant. Not only internal factors such as bone quality and quantity, cortical thickness, density of bone, but also other external factors such as implant size and technical surgery will affect the implant stability (Barikani et al. 2014).

Previous researchers mentioned that there is a high correlation between implant stability and the density of bone (Farré-pagès et al. 2011; Isoda et al. 2011; Merheb et al. 2010). However, even the significance of the statistic (p-value) in our study does not show all parameters significantly affect implant stability, among all parameters evaluated in our study showed that implant stability has a higher correlation with available space (bone height) followed by cortical thickness compared with angle insertion and density of bone.

Fig. 5.9 Progress of implant stability and density changes during the monitoring period (**a**). Stage 2, and (**b**) Stage 3

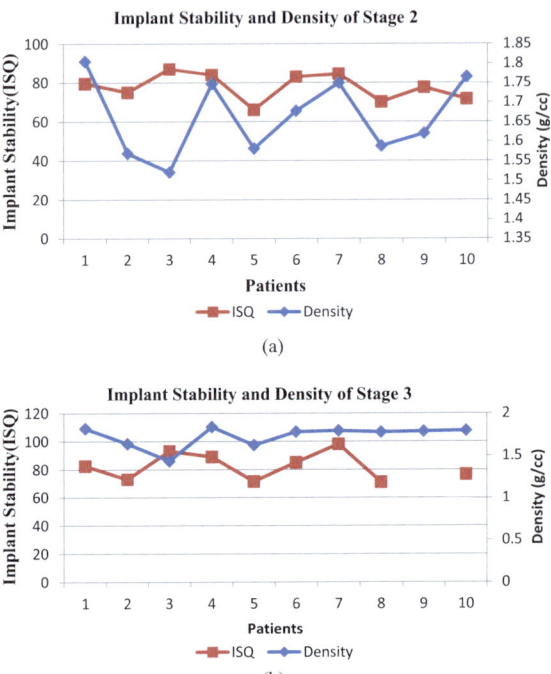

Plotting between the sizes of space availability for the implant shows that the trend of variation in the width of the mandibular is associated with the change of implant primary implant stability as shown in Fig. 5.9. The bone height has the highest correlation with primary implant stability, followed by cortical thickness.

All patients mostly underwent bone density decreasing from stage 1 to stage 2, the time between stage 1 and stage 2 is about four months, and after that the density increases from stage 2 to stage 3. The decrease in density may be related to the healing process or remodelling process of the bone post-surgery, as reported by Lin et al. (2007a, b). Mechanism of resorption of bone in the upper jaw occurs in four to six months after implant surgery and the lower jaw occurs in three to four months (Santos et al. 2002). In terms of implant stability, the implant stability of patients mostly increases from stage 1 to stage 2 and from stage 2 to stage 3. However, the increase from stage 2 to stage 3 is more significant. Increasing implant stability possibly indicates that the osseointegration has been started in this stage. If it is true, the permanent stability due to osseointegration will support the future stability of the implant.

Based on statistical analysis, in ten patients involved, there is no significant correlation between bone density with implant stability either primary implant stability or secondary implant stability. It is clear from the evidence in stage 1 to stage 2 that the density was decreasing but the implant stability in these stages was increasing. The individual correlations with other jaw parameters, and implant stability have a good relationship with cortical thickness and bone height. A patient who has a thicker cortical or bone height has the potential to have high primary implant stability. The mechanical process due to external loading during daily mastication possibly causes an increase in implant stability. Hence, the density decreasing during the healing process (between stage 1 and stage 2) does not lower implant stability. Increasing implant stability in stage 2 to stage 3 is possibly caused by an increase in density or osseointegration that was started and the mechanism of external loading of daily mastication. The decreasing and increasing density during the healing process can be detected by CBCT scanning (Kaya et al. 2012).

References

Y. Song, S. Jun, J. Kwon, Correlation between bone quality evaluated by Cone-Beam computerized tomography and implant primary stability. Int. J. Oral Maxillofac. Implants **24**(1), 59–64 (2009)

V.K. Shenoy, Considerations and pretreatment evaluation. J Interdiscip **2**, 149–157 (2012)

C.J. Goodacre, C.J. Anderson, Dental implants and Esthetics. Dentalcare.com, 3–7. Accessed: 27 January 2016 (2016)

Y. Mittal, G. Jindal, G. Sandeep, Bone manipulation procedues in dental implants.pdf. Ind. J. Dental, **7**(2), 86–94 (2016)

J.E.I.G. Brouwers, F. Lobbezoo, C.M. Visscher, D. Wismeijer, M. Naeije, Reliability and validity of the instrumental assessment of implant stability in dry human mandibles. J. Oral Rehabil. **36**(4), 279–283 (2009)

S. Ersanli, C. Karabuda, F. Beck, B. Leblebicioglu, Resonance frequency analysis of one-stage dental implant stability during the osseointegration period. J. Periodontol. **76**(7), 1066–1071 (2005)

M. Genisa, Z.A. Rajion, S. Shuib, D. Mohamad, A. Pohchi, Evaluation of stress distribution and micromotion of stress distribution and dental implant: In vivo sase study. ARPN J. Eng. Appl. Sci. **12**(4), ISSN 1819-6608 (2017)

F. Javed, H.B. Ahmed, R. Crespi, G.E. Romanos, Role of primary stability for successful Osseo integration of dental implants: factors of influence and evaluation. Int. Med. Appl. Sci. **5**(October), 162–167 (2013)

H. Barikani, S. Rashtak, S. Akbari, M.K. Fard, A. Rokn, The effect of shape, length and diameter of implants on primary stability based on resonance frequency analysis. Dental Res. J. **11**(1), 87–91 (2014)

N. Farré-pagès, M.L. Augé-castro, F. Alaejos-algarra, J. Mareque-bueno, F. Hernández-alfaro, Relation between bone density and primary implant stability. Med. Oral. Patol. Oral. Cir. Bucal. **16**(1), 62–67 (2011)

K. Isoda, Y. Ayukawa, Y. Tsukiyama, M. Sogo, Y. Matsushita, K. Koyano, Relationship between the bone density estimated by cone-beam computed tomography and the primary stability of dental implants. Clin. Oral Implants Res. 1–5 (2011)

J. Merheb, N. Van Assche, W. Coucke, R. Jacobs, I. Naert, M. Quirynen, Relationship between cortical bone thickness or computerized tomography-derived bone density values and implant stability. Clin. Oral Implant Res. **21**(6), 612–617 (2010)

D. Lin, Q. Li, W. Li, I. Ichim, M. Swain, Biomechanical evaluation of the effect of bone remodeling on dental implantation using finite element analysis. In *5th Australasian Congress on Applied Mechanics*, ACAM 2007, 10-12 December 2007, Brisbane, Australia (2007a)

D. Lin, Q. Li, W. Li, I. Ichim, M. Swain, Evaluation of dental implant induced bone remodelling by using a 2D finite element model. Australasian Congress on Appl. Mech. (December), *0–5* (2007b)

M.C.L.G. Santos, M.I. Campos, S.R. Line, Early dental implant failure: a review of the literature. Braz. J. Oral. Sci. **1**(3), 103–111 (2002)

S. Kaya, İ Yavuz, İ Uysal, Z. Akkuş, Measuring bone density in healing periapical lesions by using cone beam computed Tomography: a clinical investigation. J. Endodont. **38**(1), 28–31 (2012)

Chapter 6
Biomechanical Assessment of Dental Implant Using Finite Element Analysis (FEA)

6.1 Introduction

The relationship between mechanical stress distributions with bone remodelling after implant surgery has been reported by some researchers. Activities of remodelling around crestal bone surfaces have been identified as the main factor for the failure of the implant. However, the exact relation between stress and strain distributions with the mechanism of the remodelling process is still unclear.

Effects of repetitive loading on implants have been reported by Roberts (1998); in his theory; the effect of loading can trigger small micro cracks in the area near to implant. Those micro cracks accumulated and causing implants failure. The micro cracks are generated when the implant system loaded by fatigue loading horizontally (Huja et al. 1999). These micro-cracks have been associated with high resorption activity (Garetto et al. 1995). Micro damage caused by microcracks will increase the resorption, and hence, the quality of bone will decrease because a lot of pores are produced in the bone. Consequently, the density of bone will decrease.

This subchapter provided to analyse numerically the mechanism of stress distribution during the healing period of implant condition of pre- and post-loading through FEA analysis. Experiments are conducted in a simple model, model generated, and real in vivo data derived clinically from CBCT scanning. Evaluations are focused on the dental implant system that consists of a crown, implant, two neighbour teeth, and cortical and cancellous bone.

Various scenarios of loading are simulated to represent a daily mastication in a simple way. As an indicator, the Von Mises stress and micromotion are evaluated in each scenario. The details of the experiments are explained in the following subchapter.

6.2 Mechanism of Stress Distribution on Simple Model

Biomechanical assessments on models are performed to understand the mechanism of loading effect into implant dental system. The external loading is represented by applying a force in vertical, horizontal and torque. This force and torque are representing the real mastication process. Even though the geometry of the component of the implant system in this model is simple, this study is important as a reference before the assessment of the in vivo study is performed.

The geometry of a simple model is constructed based on the real size of each component which is measured from CBCT data. The components of the model consist of a site implant (cortical and trabecular bone), two neighbour teeth, one implant, and a crown. The geometry and sizing for each component are shown in Fig. 6.1.

Each component is assigned with the material properties as given in Table 3.3. However, all the components are regarded as a solid material which is different from actual bone which is porous naturally. This approximation is taken because of the technical limitation on how to generate the porous material model, especially for cortical and trabecular bone.

Discretization of the components into small parts or so-called meshing is performed by using a tetrahedral shape with the automatic sizing calculated by the programme. The mesh of the components is shown in Fig. 6.2.

Finite element analyses are conducted on the static structural module of ANSYS Workbench. The simulations that are performed include vertical force, horizontal force, and removal torque. All the simulations are performed to simulate the force during mastication in the pre- and post-crown conditions. As fixed supports, the bottom of cortical bone on each system is chosen as shown in Fig. 6.3.

6.2.1 Vertical Force Simulation: Pre- and Post-Crown Condition

In this simulation, a 200 Newton references vertical force is used in pre-crown and post-crown condition. The value of force is selected by referring to the range of bite force of human anatomy as reported by Fernandes et al. (2003). In the pre-crown condition, a force is applied to the top of the implant while in post-crown condition is applied to the top of the crown as shown in Fig. 6.4a. The contact surfaces between each component are regarded as bonded except for the contact between implant surfaces with cortical and trabecular. In this case, the contacts are regarded as frictional with a friction coefficient of 0.5.

Von Mises stress is measured to indicate the behaviour of stress distribution in every simulation. The stress distribution due to vertical force is shown in Fig. 6.4b. In both pre- and post-crown simulations, the maximum stresses are in the area around the neck of the implant. The value of maximum stress on post-crown condition is

6.2 Mechanism of Stress Distribution on Simple Model

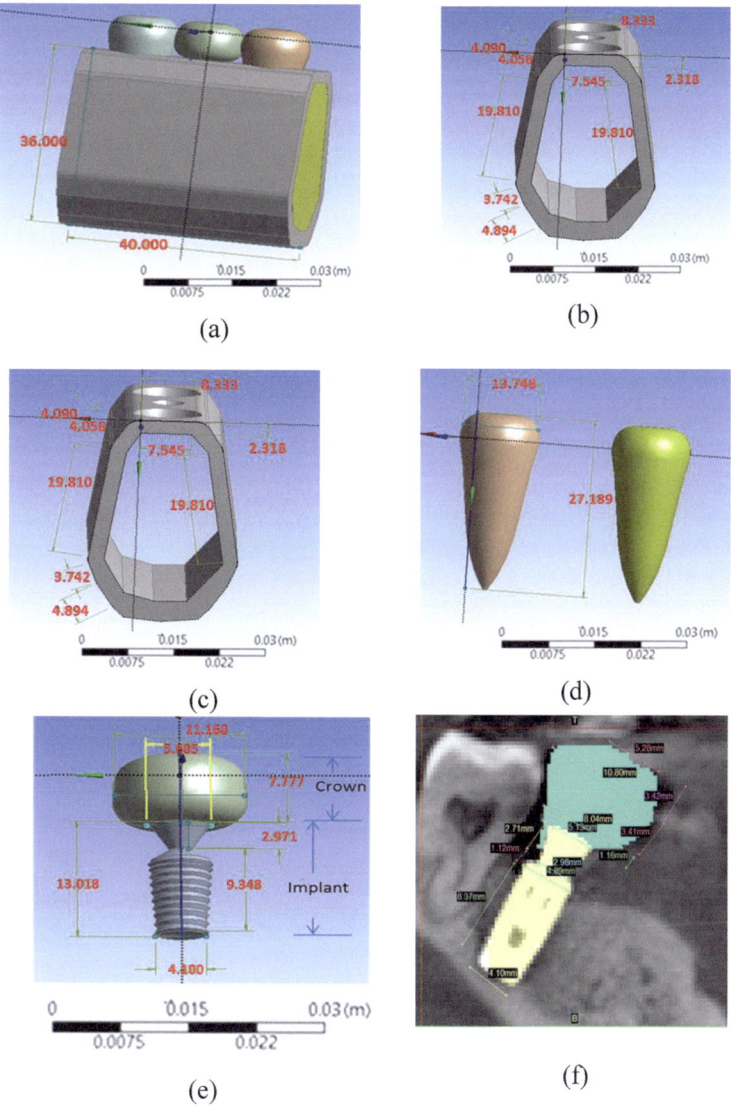

Fig. 6.1 Geometry of each component of a simple model. **a** Complete model, **b–c** geometry of cortical bone, **d** geometry of implant-crown, **e** size of implant, and **f** size of implant and crown derived from CBCT data

higher than the pre-crown condition. However, the patterns of stress distribution in the jaw in pre- and post-crown are the same.

Based on the result shown in Fig. 6.4, it can be interpreted that the mechanism of stress to reach the neighbour teeth can be explained in two ways; stress is propagated from the implant body through the cortical or from the implant body to trabecular.

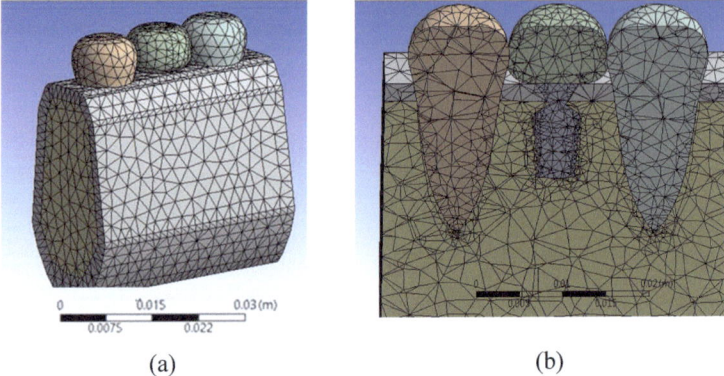

Fig. 6.2 Meshing of dental implant system **a** 3-D view, and **b** cross-section view. This simulation was solved with automatic meshing: 364,998 nodes, 255,465 mesh, and a minimum edge length was 0.341 mm

Fig. 6.3 Fixed support for Finite Element Analysis is at the bottom of the model

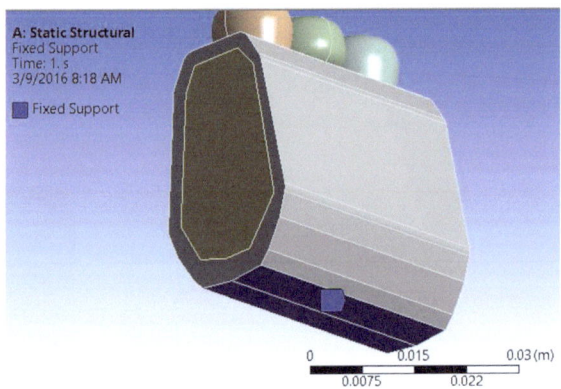

However, the distribution of stress directly from the cortical is more intense compared with the trabecular. In other words, the contribution of cortical bone to propagate the stress into neighbour teeth is more effective than trabecular.

In the neighbour teeth side (right side for single teeth), the stress on the crown condition achieves the maximum at 2 mm measured from the top of the cortical and 1 mm on the post-crown condition. In general, the neighbour teeth will receive stress due to vertical loading on pre-crown conditions which is higher than post-crown. The maximum stresses are 4.5 MPa and 1.5 MPa in pre- and post-crown conditions respectively, occurring at around 2 mm measured from the top of cortical. Below 3 mm, the stress is decreasing to the lower part, and the Von Mises stress for this level is low enough, below 0.5 MPa. In the neighbour teeth, the stress due to vertical loading is only 11%-25% from the contact area of the implant.

6.2 Mechanism of Stress Distribution on Simple Model

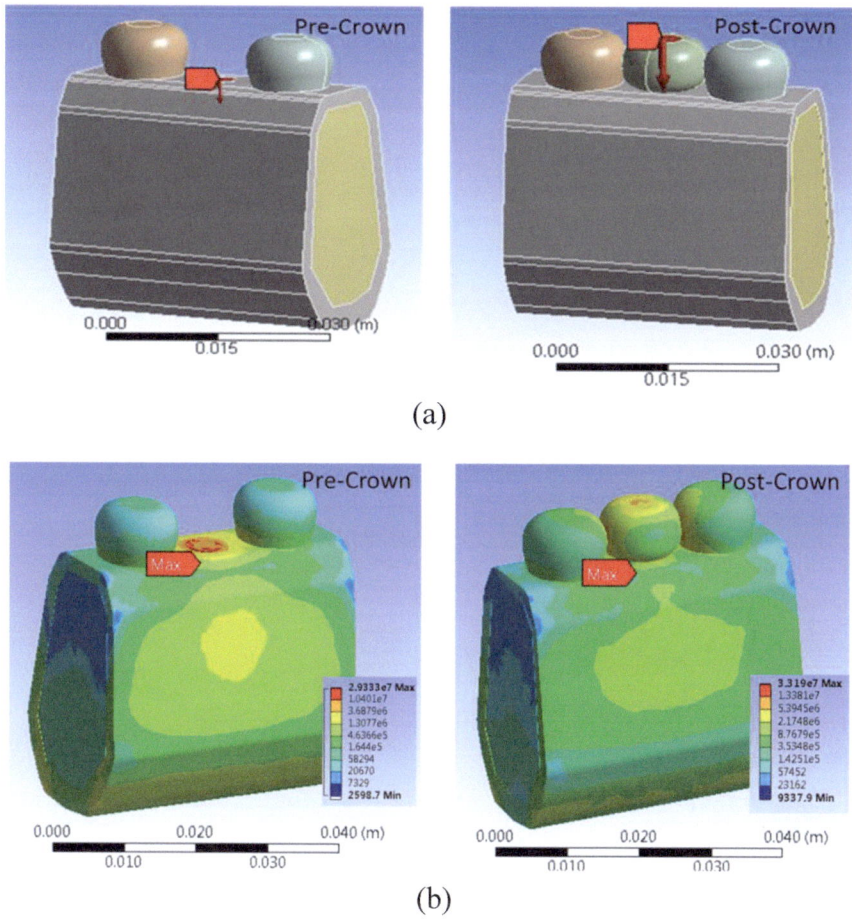

Fig. 6.4 **a** Push-out simulation with a 200 N vertical force is loaded into the implant dental system, pre-crown (left) and post-crown conditions (right), and **b** Von Mises stress distribution in 3-D view resulted from FEA simulation

6.2.2 Horizontal Force Simulation: Pre- and Post-Crown

Horizontal force simulations are performed to replicate a 200 Newton of horizontal force (from palatal to buccal direction) as shown in Fig. 6.5a. In the pre-crown, the contact area of the force is located at the top of the implant body, while in the post-crown, the force is located on buccal side of the crown. The fixed supports for both simulations are on the bottom side of the jawbone. The meshing of this model is generated the same as the previous study on vertical loading, and the difference is only in the direction of the force.

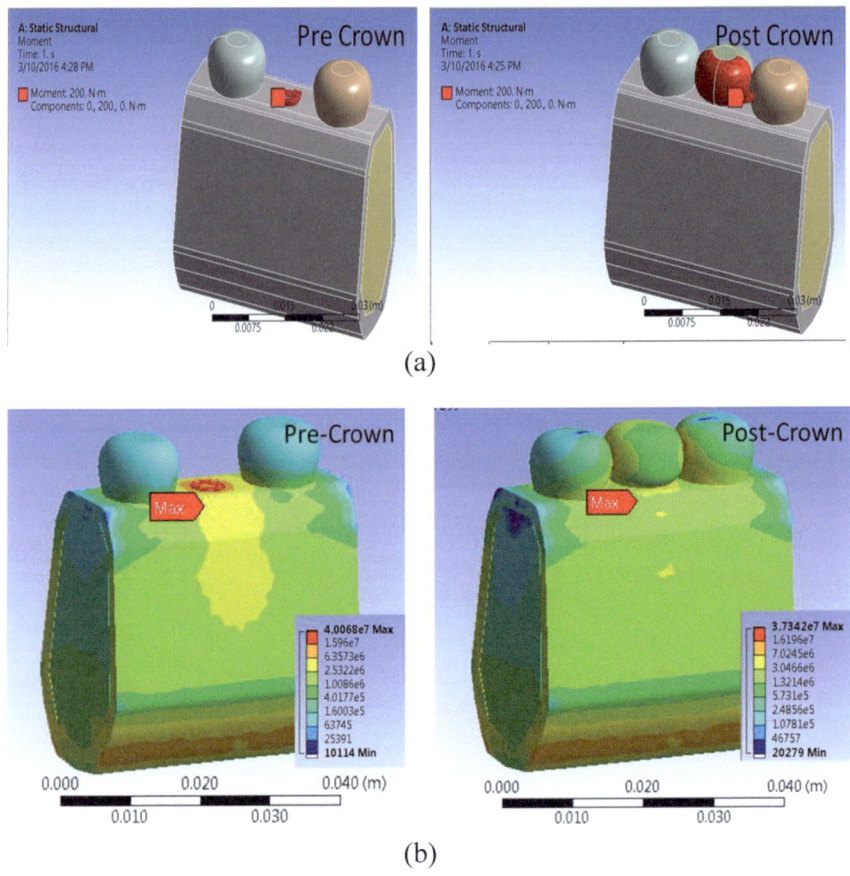

Fig. 6.5 Horizontal force simulation of pre- and post-crown, a). location of horizontal force (200 Newton), and b). Von Mises stress distribution in 3-D view of pre- and post-crown

Figure 6.5b shows the Von Mises stress in a 3-D view because of the simulation. The stress distribution in pre-crown conditions is propagated with the high concentration in the area close to the implant body. Meanwhile, in the post-crown condition, the distribution of the Von Mises stress is propagated radially along the whole area of the jawbone. It shows that the propagation of the stress is easier transmitted through the cortical than the implant body itself, and hence, the distribution of the stress in the outer part of the jawbone is more homogenous. The maximum stress of pre-crown is in the top of cortical bone, while in the post-crown, the maximum stress is located in the contact area between crown and neighbour teeth. The maximum stress on post-crown is 37.34 MPa which is 6.67% lower than the crown condition (40.01 MPa).

In general, up to 2 mm, the stress resulting from pre-crown is dramatically very high compared to the pre-crown condition. Understandably, the pre-crown condition

6.2 Mechanism of Stress Distribution on Simple Model 73

underwent higher stress because the force is pointed directly to the implant body. In this case, the crown has protected the implant by reducing the stress.

6.2.3 Removal Torque Simulation

Removal torque simulation is performed to understand the responses of the dental system on a torque/rotation force. In natural masticatory, the position of foods during masticatory can also generate a torque that makes a dental implant become rotated. In this study, a 200 Nm torque is directed into the top of the implant and crown as pre- and post-crown condition as shown in Fig. 6.6a. Finite element analysis is performed by using parameters that are similar to the previous simulation on vertical and horizontal simulation except for the loading.

The results are shown in Fig. 6.6b. The maximum Von Mises stresses are in the area where the torque is located. The pattern of stress distribution due to removal torque in both pre- and post-crown conditions is the same, with the stress more concentrated in the area surrounding the implant.

The difference between generated stress at pre- and post-crown is quite big on the implant side compared to the neighbour teeth location. It shows that the removal torque has a strong effect on the implant body compared to the neighbour teeth. Stress propagated in the neighbour teeth is not too much different between pre- and post-crown conditions due to removal torque.

6.2.4 Summary of Result

Experiments on pre- and post-crown by using three different loadings which consist of vertical loading, horizontal loading,, and torque on dental implant systems have been conducted. The objective of this experiment is to understand the mechanism of stress distribution as a response to external loading. The Von Mises stress has been chosen to represent the stress distribution. Digitations on the results are performed by putting the measurement probe of Von Mises stress at a certain location of interest. For each experiment, the stress distribution between pre- and post-crown conditions has been compared and plotted.

Based on the results, it can be summarized that: In three different external loading, cortical plays important roles in propagating the stress in the dental implant system. The cortical can protect the trabecular from high transfer of stress. In all cases, at the same level of measurement, the stress in the trabecular is lower than cortical. Compared to other external loadings, the external loading due to torque can produce a lot of stress at the top of the cortical and it was propagated into the whole area of the jawbone. The high-intensity area due to this loading is wider compared with other external loading. It means that the uncomfortable (destruction) zone of bone around the implant due to the torque is wider compared with other external loading. The

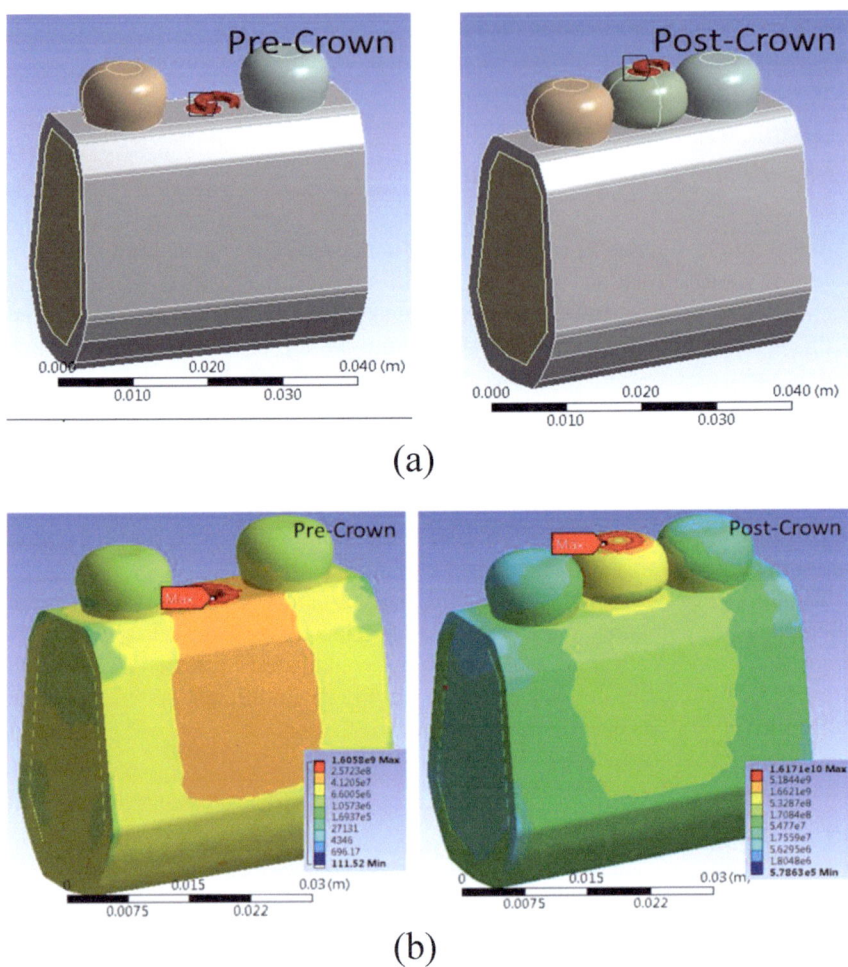

Fig. 6.6 Removal torque simulation for pre- and post-crown: **a** Location of removal torque, and **b** stress distribution in 3-D view

potential of implant failure or low osseointegration of implant/low implant stability is higher if too much torque is subjected to a dental implant.

6.3 Effect of Cortical Thickness and Friction Coefficient on Stress Distribution

The primary and secondary implant stability is determined by some important factors such as the quantity and quality of bone, the technique of how the implant is inserted, and the environment of the mouth during the recovery process (Annibali et al. 2008). The stress distribution during loading will be determined by the stability of the implant and its mechanism which depends on internal and external variables.

In this research, two variables, the cortical thickness and frictional coefficients, are evaluated to understand the effects of it on the stress distribution on the dental implant system. In this experiment, different models of dental systems consisting of cortical, trabecular, body implant, crown, and two neighbour teeth are used. These models are constructed from 3-D images of CBCT of dental implant patients.

6.3.1 Effect of Cortical Thickness on Stress Distribution

To investigate the effect of cortical thickness on stress distribution, models with different cortical thicknesses (2.30, 2.85, 3.53, and 3.93 mm) were developed. Those models are generated to approximate the regular increment for every 2 mm. However, it is difficult technically to get the linear increment in developing the model of cortical thickness. Hence, the generated model is developed based on increment of voxel. In this experiment, only cortical are varied, while other components such as neighbour teeth, implants, and crowns are maintained the same. The dental implant systems for different cortical thicknesses are shown in Fig. 6.7. The different cortical thicknesses are obtained by modifying the cortical during segmentation by using morphology operation tools on MIMICS software. The original cortical mask has been modified by dilating/shrinking the pixel.

Each model is discretized by using similar parameters; hence, the size and density of the mesh are almost similar in every model. Material assignment is performed by using the value for each component as given in Table 3.3. In this experiment, a 200 Newton vertical force as in the previous simulation of a simple model is used in pre- and post-crown conditions. Fixed support is at the bottom of the model; hence, there are no different parameters for each model that are used, except the cortical thickness.

Finite element analysis is performed with the same parameters on each model. The properties of contacts between components are approximated with the frictional contact and bonded contact. Frictional contact is the contact between the implant with cortical and trabecular, and bonded contact is the contact between the crown to implant, and teeth to the bone either cortical or trabecular. The stress distributions of because of the simulation are shown in Fig. 6.8.

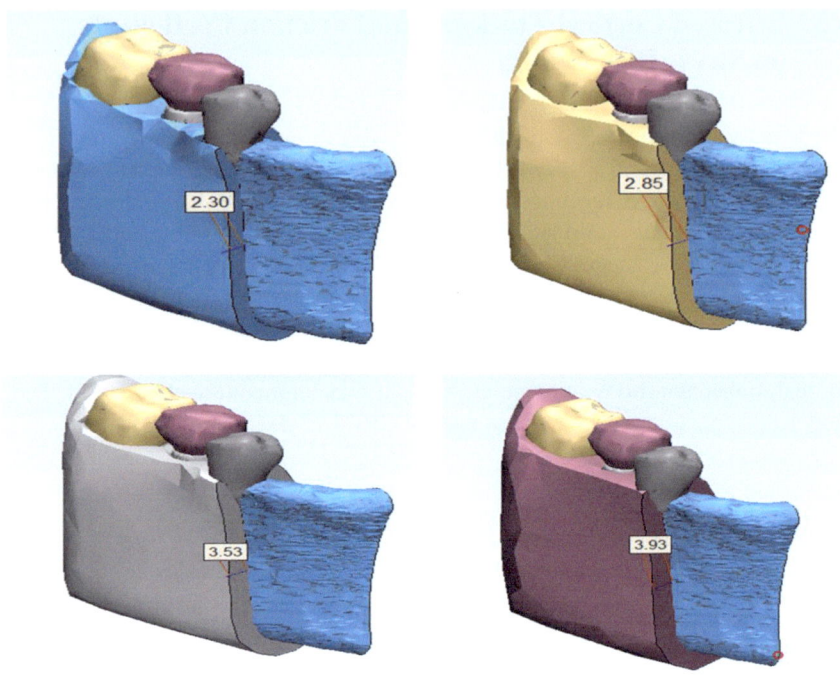

Fig. 6.7 Models of dental implant systems with different cortical thicknesses of 2.30, 2.85, 3.53, and 3.93 mm

6.3.2 Effect of Friction Coefficient on Stress Distribution and Micro Motion

To investigate the effect of friction coefficient on stress distribution, a model in previous experiments is used by modifying the contact behaviour between implant and bone (trabecular and cortical). In this contact area, the friction coefficient is set to 0.2, 0.3, 0.4, and 0.5. In every simulation, all the parameters including meshing, fixed support, location of probe measurement, and surface contact behaviour are kept constant, and only the friction coefficient is different.

The stress distribution has an impact on the static friction at the contact surface as reported by Maegawa et al. (2016). However, the friction coefficient whether affects the stress distribution is still not investigated. In this experiment, we investigated that process, the effect of friction coefficient on the stress distribution.

Figure 6.9 shows that on each position of measurement, the Von Mises stress level is changed with the friction coefficient. Increasing in friction coefficient is followed by decreasing in the Von Mises stress level. The friction coefficient which represents the quality of the surface contact between two objects affects the stress propagation and enhances the distribution of the stress as well. Increasing in friction coefficient

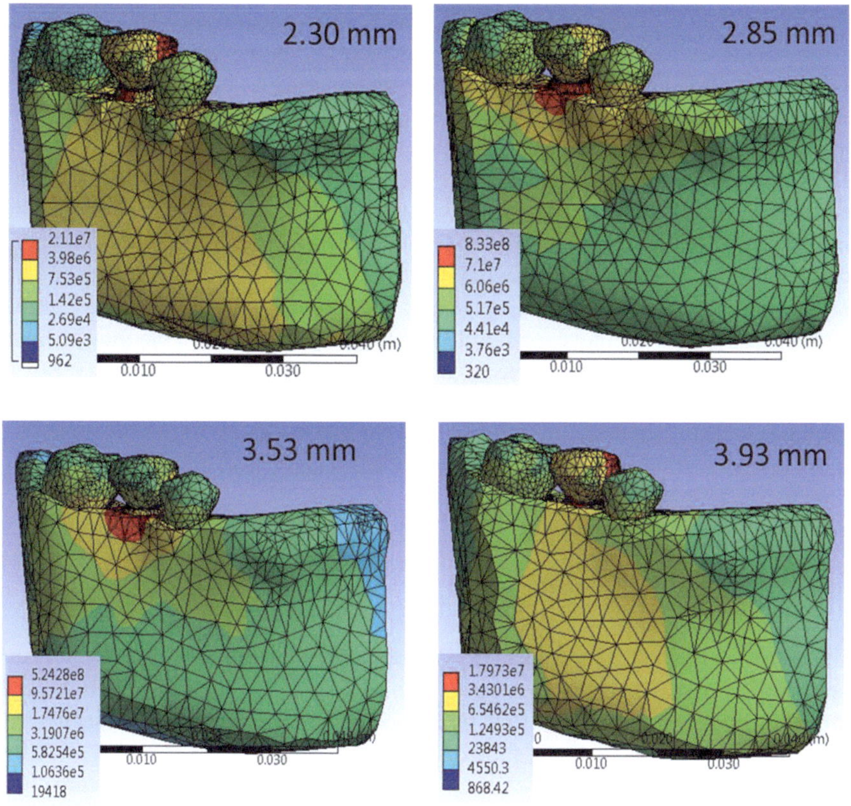

6.3 Effect of Cortical Thickness and Friction Coefficient on Stress Distribution

Fig. 6.8 Von Mises stress in a 3-D view of each model with different cortical thicknesses

will give more chance for the stress to be transferred to the other side of different objects, and hence, the number of transferred stresses is reduced.

In the dental implant system, the loading not only produces a stress distribution but also can produce micro movement of components which may occur at two interfaces, the implant-abutment connection and the bone-implant interface (Karl et al. 2015).

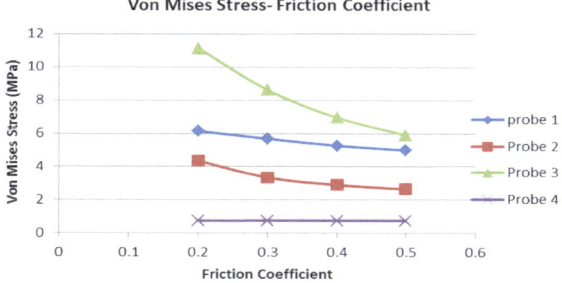

Fig. 6.9 Von Mises stress at different probes for different friction coefficients

However, the case of engineering issues which may oaccure, such as loosening of the abutment screw is not considered in this simulation. In this experiment, the micro motion is regarded as a movement between the bone-implant interfaces only. In the FEA study, micro motion is measured by calculating the displacement difference between reference points with the target/object.

The friction coefficient contributes to the micromotion of the dental body. This experiment shows that if the friction coefficient at the contact surface between the implant and bone is increased, then the possibility of the implant body moving is low. In other words, the friction coefficient can reduce the micromotion of the dental implant body due to external loading such as in the mastication process. The external loading also affects the movement of neighbour teeth; fortunately, the movement of neighbour teeth is very low. The inserted implant by the immediate implant protocol does not jeopardize the existing natural teeth during the mastication process. In this case, a 200 N vertical force is still safe for the implant because the generated micromotion is still in the tolerance range (50–150 µm) (Karl et al. 2015).

6.3.3 Summary of Result

Clinically, the test to measure the effect of external loading on stress distribution and micromotion during mastication directly is mostly impossible due to the limitation of the availability of the sensor. Like other mechanical problems, the biomechanical assessment on implant dentistry, the FEA is a tool that can be considered to solve the problem mentioned before, if the parameters that are used during analysis are close enough to the real condition.

The stress due to the external loading will propagate faster in the cortical trabecular. Therefore, the generated stress around the neighbour teeth mostly comes from the upper cortical rather than the implant body and trabecular. In general, increasing the cortical thickness will increase the stress transfer into the surrounding area in either a horizontal or vertical direction. Because the propagation of stress is high in the cortical, the ability to transfer the stress through the cortical is higher. This condition can protect the bone from suffering from high stress due to loading, which means that it is good for clinical purposes.

Inversely, the friction coefficients decrease the ability to transfer the stress. Increasing in friction coefficient will decrease the transferred stress. Therefore, the dental implant with a high friction coefficient will protect the bone from overloading the stress due to external loading. It is possible that in the high friction coefficient, the modelling/remodelling will grow faster, and the osseointegration will be achieved early.

The measurement of micromotion on the implant and surrounding teeth showed that the external loading can cause the micro motion in the dental which is larger than neighbour teeth. It is understandable because the coupling between implant bones is still not optimum compared with the teeth and bone, especially for patients with delays in the healing process. It corresponds with the experiment that increasing in friction coefficient can reduce the micro motion in the dental implant.

References

S. Annibali, M. Ripari, G. LA Monaca, F. Tonoli, M.P. Cristalli (2008). Local complications in dental implant surgery: prevention and treatment. ORAL & Implantology, **1**(1), 21–33

C.P. Fernandes, P.O.J. Glantz, S.A. Svensson, A. Bergmark, A novel sensor for bite force determinations. Dent. Mater. **19**(2), 118–126 (2003)

L.P. Garetto, J. Chen, J.A. Parr, W.E. Roberts, Remodeling dynamics of bone supporting rigidly fixed titanium implants: A histomorphometric comparison in four species including humans. Implant Dent. **4**, 235–243 (1995)

S.S. Huja, T.R. Katona, D.B. Burr, L.P. Garetto, W.E. Roberts, Microdamage adjacent to endosseous implants. Bone **25**(2), 217–222 (1999)

M. Karl, F. Graef, W. Winter, Determination of micromotion at the implant bone interface: an in-vitro methodologic study. Dentistry **5**(4), 1–5 (2015)

W.E. Roberts, Bone tissue interface. J. Dent. Educ. **52**, 804–809 (1998)

S. Maegawa, F. Itoigawa, T. Nakamura, Effects of stress distribution at the contact interface on static friction force: numerical simulation and model experiment. Tribol. Lett. **62**

Chapter 7
Biomechanical Assessment of Patient with High Implant Stability

7.1 Biomechanical Assessment of Patients with High Implant Stability

Experiments of biomechanical assessment on patients with high dental implant stability have been conducted. A dataset of female patients (42 years old) with normal edentulous molar cases was used. All dental implant treatments are conducted with delayed implant procedures which are conducted by specialists in Hospital USM. The dental implant was placed in a MegaGen Implant with the size 4 mm × 10 mm which is installed into the molar of the mandible. Implant stability of her implant was monitored by using RFA measurement. She has high primary implant stability (82 ISQ scale), and her secondary implant stability was increased during the period of monitoring (87 and 93 ISQ after 3 months installation and 4 months after installation respectively).

A biomechanical assessment of her implant system is evaluated around the implant site and the two nearest teeth. The implant system that is selected consists of: cortical bone, trabecular bone, body implant, crown, molar 2 tooth, and premolar 2 tooth as shown in Fig. 7.1. To represent a high implant stability condition, the contact surface between implant surfaces and bone is regarded as frictional with a friction coefficient of 0.5. The geometries of the implant system model are derived from raw CBCT data.

7.1.1 Behaviour of Stress Distribution on High Implant Stability Patient

FEA simulations consist of vertical force, horizontal force, and removal torque simulations which are conducted to know the distribution of the stress in the dental system in the pre- and post-crown condition. The result of this simulation is shown in Fig. 7.2

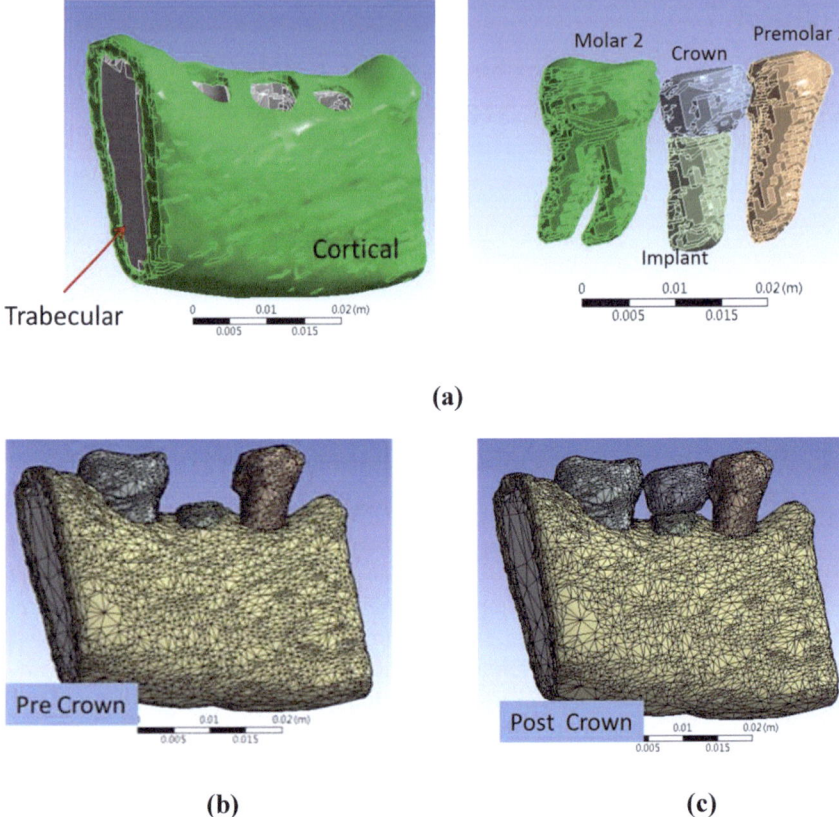

Fig. 7.1 Model of dental implant system derived from CBCT of the patient with high implant stability. **a** Components of the dental implant system consist of cortical and trabecular of the jawbone, two neighbour teeth, implant body, and crown, **b** pre-crown model, and **c** post-crown model

where the Von Mises stress is used to represent the stress distribution with stress intensity colour-coded.

In high implant stability patients, all the maximum stress is at the top of the cortical at the surrounding implant. The stress distributions in pre- and post-crown are almost similar, especially in the horizontal direction. However, from up to bottom, the spreading of stress between pre and post is different, and the stress distribution of pre-crown is wider than post-crown. In other words, in the pre-crown where the masticatory force is applied directly to the top of the implant, the stress distribution is transferred into the surrounding bone of the jaw. From that figure, the areas affected by the stress in the crown are wider than the post-crown.

These results show that the stress is propagated more progressively in the vertical than lateral direction, and the areas affected by stress in the crown are wider than post-crown. The installation of the crown can reduce the generated stress in the jaw

7.1 Biomechanical Assessment of Patients with High Implant Stability

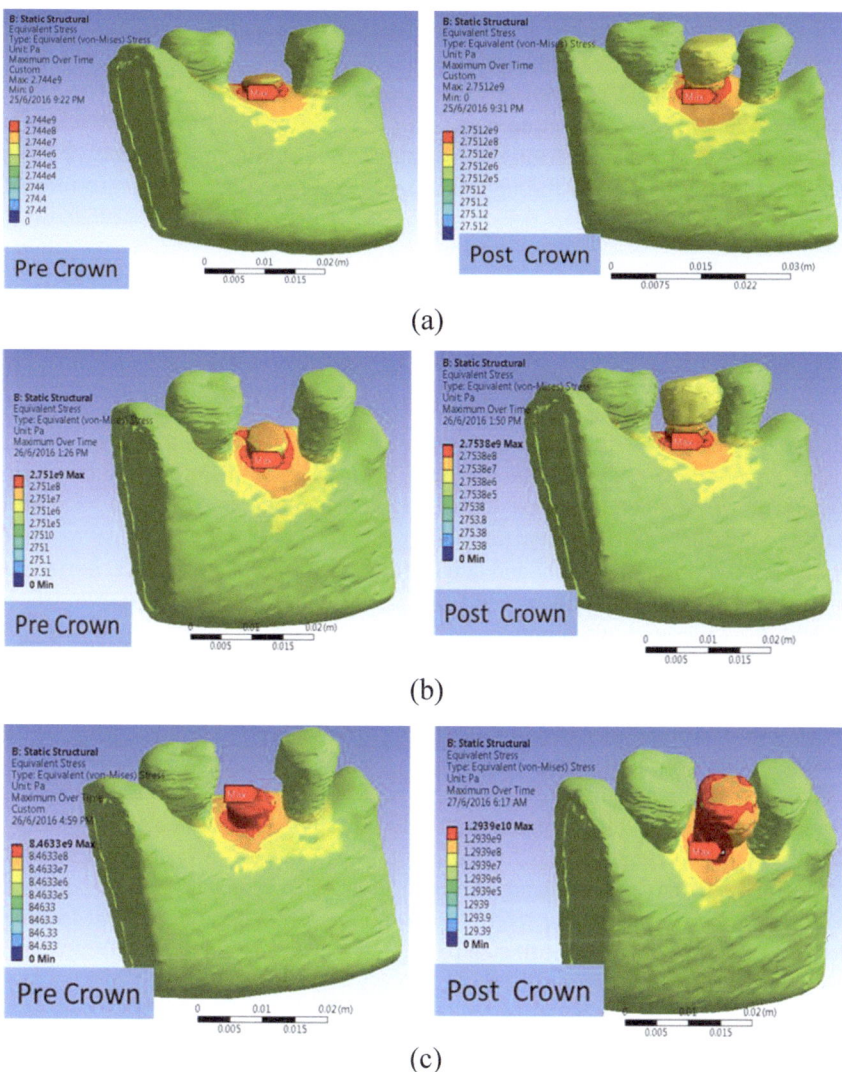

Fig. 7.2 FEA simulations for different types of loading on pre- and post-crown conditions **a** Vertical force, **b** horizontal force, and **c** removal torque simulation

system. As an example, in the pre-crown condition of horizontal force, the maximum Von Mises stress on the implant body is about 120 MPa, while in the post-crown condition is about 63 MPa, and stresses are reduced more than 50%. The distribution of the stress in the jaw system between horizontal and vertical directions for pre- and post-crown conditions is distributed similarly in the radial direction.

When a loading is directed vertically or horizontally, the generated stress distribution is almost similar. The stresses are distributed mostly around the implant from up to the bottom of the implant. The spreading of stresses is more concentrated in the upper part of the jawbone (in the upper cortical). In vertical and horizontal loading, the second premolar receives more effect than the second molar tooth. The generated stress due to removal torque is stronger than vertical and horizontal force. In this case, stress propagated from the implant body to the nearest neighbour teeth. The propagation is not only through the upper part of the cortical but also through the trabecular bone.

7.1.2 Micro Motion of Implant and Neighbour Teeth

Not only does it cause stress, which is distributed along the dental implant system, but also the loading can introduce the movement (micromotion) of the dental implant and neighbour teeth relative to the jawbone. To investigate the micromotion of each component of the dental implant during different loading, some probes have been placed to measure this displacement.

The micromotion of implant and neighbour teeth due to vertical force, horizontal force, and removal torque at pre and post conditions are shown in Table 7.1. In general, the generated micromotion of the implant at pre-crown condition is higher than post-crown. The generated micro motion of dental implant during vertical and horizontal loading ranges from 21 to 29 μm and removal torque ranges from 210 to 220 μm. The smallest effect of crown installation on micromotion occurs during horizontal loading.

The micromotion of neighbour teeth, second premolar, and second molar in all types of loading are small compared to the micro-motion of the implant. Micromotion of these neighbour teeth ranges from 5 μm to 8 μm during vertical and horizontal loading and 3 μm to 29 μm during removal torque. Crown installations do not affect the micromotion of neighbour teeth, especially on vertical and horizontal loading. However, the removal torque affected neighbour teeth significantly.

Table 7.1 Micro motion of implant system due to a vertical force, horizontal force, and removal torque of high implant stability patient

Micromotion (μm)	Simulation					
	Vertical force		Horizontal force		Removal torque	
	Pre	Post	Pre	Post	Pre	Post
Implant	28.61	21.24	21.03	22.32	211.12	219.10
Premolar	5.27	5.49	5.43	5.40	10.70	2.88
Molar	6.99	7.28	6.99	6.76	22.82	15.53

7.2 Summary of Result

Biomechanical assessments of patients with high implant stability have been conducted in the different loading types at pre- and post-crown scenarios. Evaluations included the stress distribution analysis and micro motion estimation on the dental system.

In all scenarios of experiments, the upper part of cortical injury had the highest stress compared with other places. From this area, the stress is propagated in all directions, vertically and horizontally, and can reach the neighbour teeth. The removal torque produced the highest stress in the upper cortical (up to 1500 MPa); however, the horizontal and vertical force can produce the stress up to 600 MPa.

The neighbour teeth are significantly affected only by removal torque. The Von Mises stress of neighbour teeth due to removal torque can be achieved up to 200 MPa in the upper cortical and up to 40 MPa in the lower cortical and trabecular. Compared to vertical and horizontal force, stress resulting from removal torque is very high. Vertical and horizontal force can generate stress in the upper cortical of neighbour teeth up to 45 MPa and lower cortical and trabecular is about 5 MPa. These experiments showed that the removal torque is more dangerous for implant stability rather than vertical force or horizontal force. In addition, measurement of micro motion showed that the micromotion due to removal torque is larger than vertical and horizontal force. The micromotion due to removal torque goes up to 220 μm, while the vertical and horizontal force is about 20–30 μm.

The osseointegration can be achieved earlier if the environment for bone to grow normally can be maintained. Therefore, patients who undergo the removal of torques extensively, especially in early implant insertion, their bone receives more stress that exceeds the limit the harmless for the bone to grow and the implant will have more movement. It is possible that the osseointegration is difficult to achieve early for this condition.

7.3 Biomechanical Assessment of Patient with Moderate Implant Stability

The stress mechanism on moderate implant stability was investigated on a model derived from CBCT data of patients who have moderate implant stability. A dental implant patient (a male, 25 years old) who has a normal edentulous on his molar 1 of mandible was selected. Immediately after implant surgery, his implant stability was measured. His primary implant stability was 73 on the ISQ scale and after three months and four months after surgery; his implant stability was 80 and 83 on the ISQ scale, respectively. In general, his implant stability was increasing progressively even though the increment was not so high.

The model of the dental implant system obtained from CBCT data consists of an implant, crown, two neighbour teeth (second premolar and second molar), and

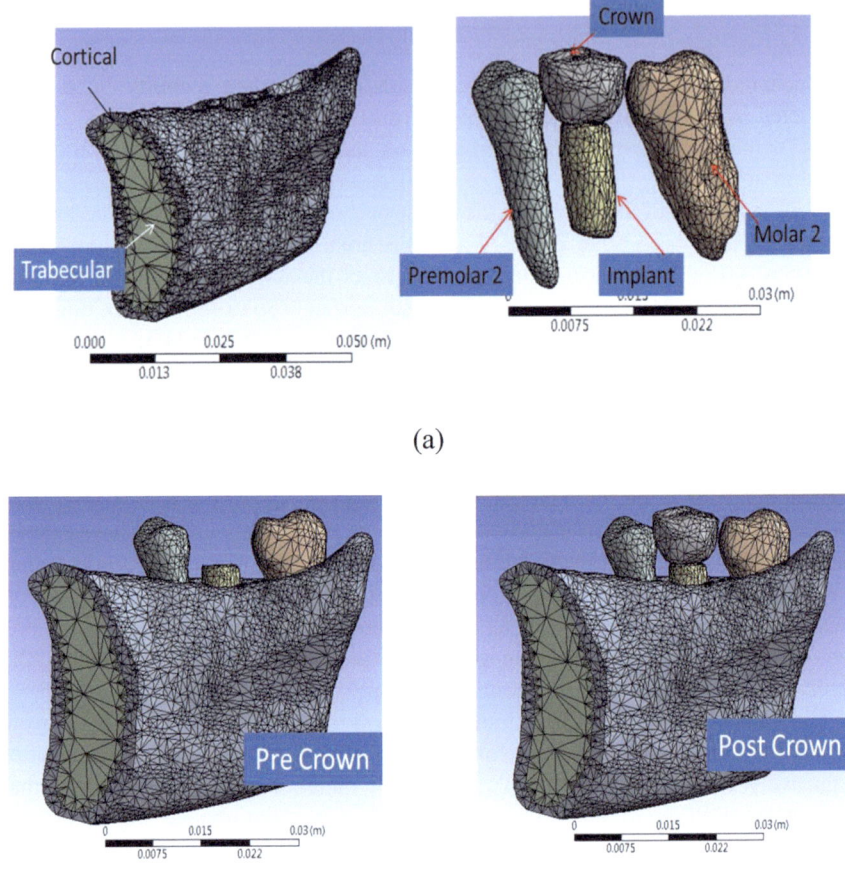

Fig. 7.3 Geometry of pre- and post-crown of patients with moderate implant stability, **a** components of the dental implant system, and **b** complete meshing of the dental implant system for pre- and post-crown conditions (automatic meshing: 24,384 nodes, 91,819 elements)

cortical and trabecular bone. Those components of the dental implant model and the condition of pre- and post-crown are shown in Fig. 7.3.

7.3.1 Behaviour of Stress Distribution on Moderate Implant Stability Patient

Stress distribution behaviour during different types of loading on a model of moderate implant stability was conducted. The simulation consists of vertical force, horizontal force, and removal torque for pre- and post-crown conditions. In each scenario, the parameters such as geometry, material properties, meshing, fixed support, and boundary conditions are maintained constant. Only the location and direction of loading are different. To represent a moderate implant stability condition, the contact surface between implant surfaces and bone is regarded as frictional with a friction coefficient of 0.4. Generated Von Mises stresses from three types of loading in pre- and post-crown conditions are presented in Fig. 7.4.

In general, the loading process on a model of moderate implant stability produced the stresses that are concentrated in the upper part of the cortical. For example, the stress resulting from vertical force, the stress can be more than 50 MPa either in pre- or post-crown conditions. While in the lower part, especially in the trabecular, even in the area around the implant, the stress is still lower than 10 MPa. The stress received by neighbour teeth is also low (less than 10 MPa). Compared between the three types of loading, the stress due to vertical force is lowest and removal torque produced the highest stress, especially in the upper part of cortical.

7.3.2 Micro Motion on Moderate Implant Stability Patient

Micro motions of implant and neighbour teeth of patients with moderate implant stability have been measured for different types of loading. Micromotion of this component was measured relative to the jawbone. The measured micro motion of implants and neighbour teeth on different loading at pre- and post-crown are summarized in Table 7.2.

Micro motions are evaluated at implant, second premolar, and second molar by comparing the deformation of its component with the deformation of the reference point. In all simulations, the micromotion of the implant at post-crown condition is smaller than pre-crown. Maximum micromotion occurred when removal torque was applied at pre-crown condition, and the micro-motion at this condition was 124 μm. The premolar and molar teeth have a small movement that is less than 2 μm except on the post-crown. High micro motion in the implant is related to the high intensity of stress in this area.

Evaluation of micro motions on dental implants and neighbour teeth shows that the horizontal forces are not causing a significant movement. The micromotion of the implant is about 9–17 μm while in premolar and molar is less than 1 μm.

7.4 Summary of Result

Biomechanical assessments of patients with moderate implant stability have been assessed by including stress distribution and micro motion analysis. Three types of loading have been simulated, and they are: vertical force, horizontal force, and removal torque. The simulations are performed in pre- and post-crown conditions.

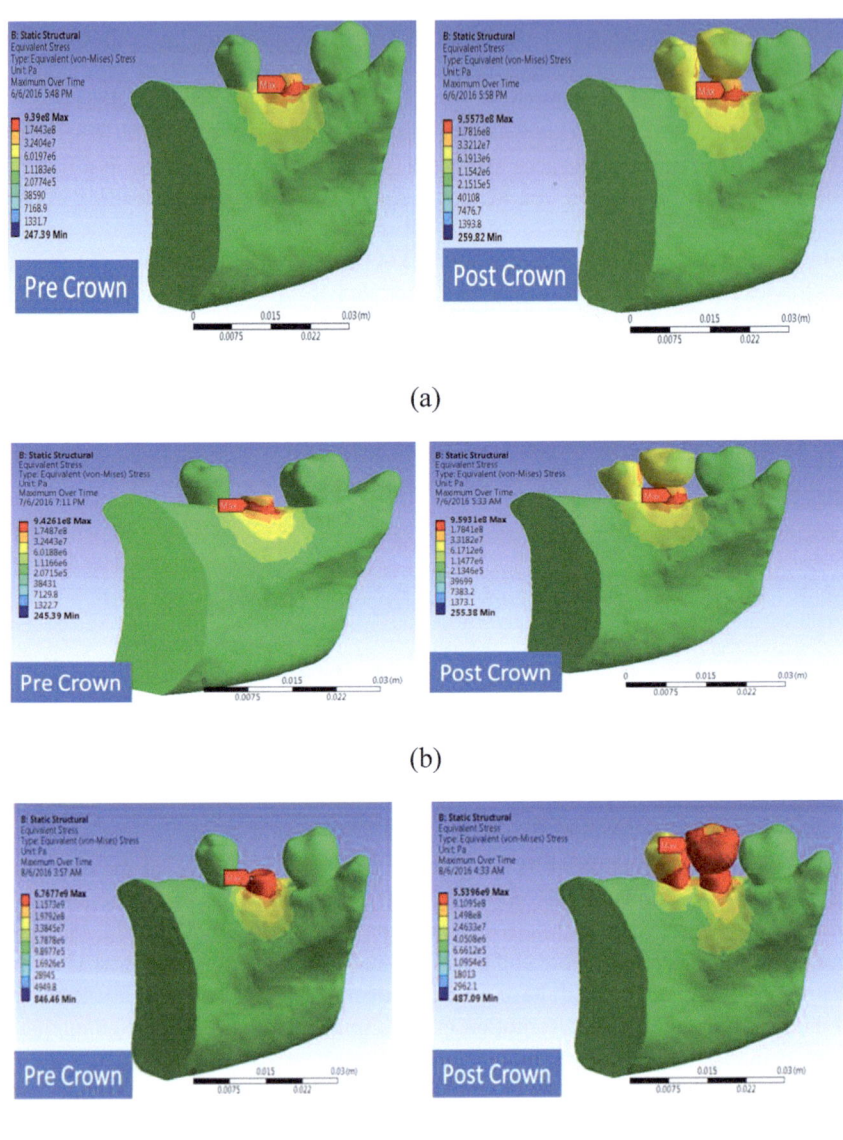

Fig. 7.4 **a** Von Mises stresses on moderate implant stability model generated from different types of loading. **a** Vertical force, **b** horizontal force, and **c** removal torque simulation at pre- and post-crown conditions

7.4 Summary of Result

Table 7.2 Micromotion of the implant system of the patient with moderate implant stability

Micro motion(μm)	Simulation					
	Vertical force		Horizontal force		Removal torque	
	Pre	Post	Pre	Post	Pre	Post
Implant	19.40	18.80	16.82	9.42	124.16	45.48
Premolar 2	0.43	0.69	0.20	0.35	0.10	12.25
Molar 2	0.05	0.17	0.56	0.26	1.18	0.67

In vertical force loading, the results showed that only the upper part of the cortical was affected and received high stress. In both pre- and post-crown conditions, the higher stress area is located up to 2 mm measured from the top of the cortical. For the rest (in the low part), the generated stress is low which is lower than 10 MPa. This generated stress is still in the range for bone to grow normally. The micro-motion generated by this loading is also not too significant, the micro-motion on the implant itself is only about 20 μm and in the neighbour teeth is very low (less than 1 μm).

In the simulation of horizontal loading, where the 200 N force is applied on the pre- and post-crown, the generated stress in the jawbone is not so different between pre and post. The trend and intensity of stress are almost like each other. As like in vertical loading, horizontal loading can only affect the upper part of the cortical, below 2 mm, the stress is in the range for bone to grow normally. The micro motions that are generated during horizontal loading also do not cause a high micro motion either on dental or on neighbour teeth. The horizontal loading still produces a safe range of stress for bone to grow normally.

Removal torque gives some different results on stress distribution and micromotion. When the torque is pointed directly into the top of the implant body, the stress can produce more stress in the upper part of the cortical. Not only generating high stress but also the removal torque caused a bigger micro motion. The micro-motion on the dental implant can vary from 45 μm up to 125 μm in the post- and pre-crown condition, respectively. However, the micro-motion of neighbour teeth is not so high except on the premolar of post-crown condition.

Chapter 8
Biomechanical Assessment on Patient with Low Implant Stability

8.1 Biomechanical Assessment on Patient with Low Implant Stability

Biomechanical assessment of a patient who has low implant stability is conducted on a model generated from a female patient 38 years old who has a second molar edentulous problem. Implant surgery was conducted to replace her second molar tooth. The size of that implant is 4.5 mm (diameter) x 10 mm (length). All the procedures of implant treatment and monitoring are the same as for other patients. Her implant stability during the monitoring period was assessed by using an RFA instrument. This patient has low primary implant stability (66 on the ISQ scale) and increase to 75 after 4 months and 73 after 5 months since implant surgery. The record showed that the progress of her implant stability was not so progressive. This condition is interesting to be analysed in more detail in the FEA study to understand the mechanism of stress distribution during pre- and post-condition.

Biomechanics assessment of her implant system during the loading stage has been conducted numerically by using FEA. To perform this analysis, the geometry of the dental implant system was derived from her CBCT data. The model of the dental implant system of this patient consists of trabecular and cortical bone, implant and crown, and two neighbour teeth. The components of the dental implant system of this patient are shown in Figure 8.1.

Loading as representative of force during the mastication process consists of vertical force, horizontal force, and removal torque. Simulations are performed in pre- and post-Crown conditions; all parameters used in each simulation are kept the same. Hence, the result of the analysis does not depend on the parameter but only on components of the dental implant system either pre- or post-crown condition.

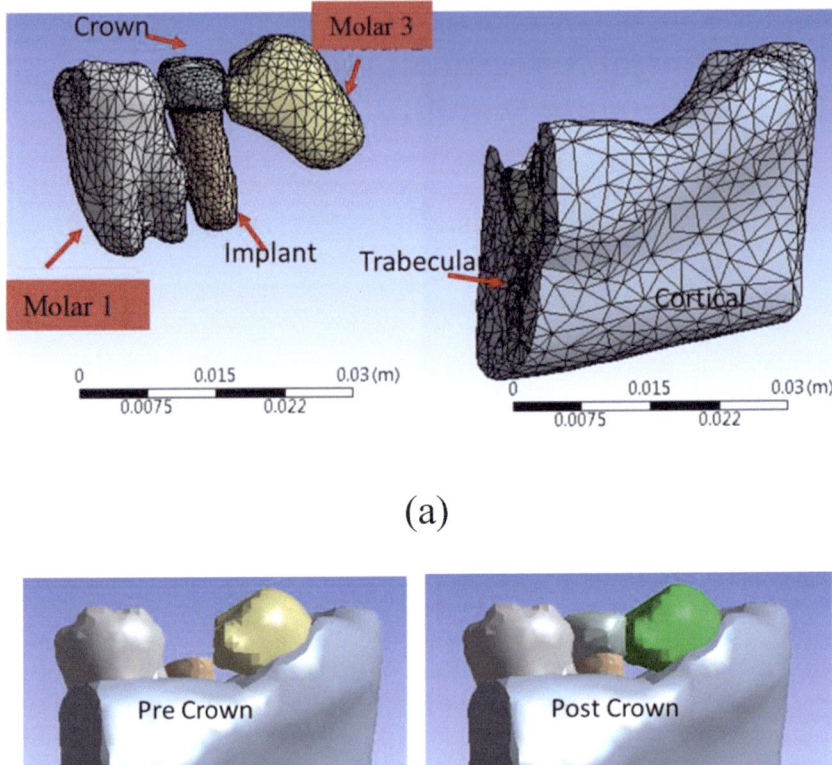

Fig. 8.1 Model of dental implant system of a patient with low implant stability. **a** Components of the dental implant system, and **b** pre- and post-crown model (automatic meshing: 5655 nodes, 17,856 elements)

8.1.1 Behaviour of Stress Distribution on Low Implant Stability Patient

FEA simulations of loading on pre- and post-crown conditions are performed by assuming that the contact surface between implant and bone is a frictional surface with a friction coefficient was 0.2. Von Mises stress was selected to represent the

8.1 Biomechanical Assessment on Patient with Low Implant Stability

generated stress distribution in each simulation. The stress distribution for each vertical loading, horizontal loading, and removal torque is presented in Figure 8.2.

The installation of the crown on the implant makes the molar tooth contact with the dental implant, which affects the stress propagation strongly. In vertical loading (Figure 8.2a), the generated stress is very strong in the area around the implant body. The maximum stress in the pre-crown simulation is about 307 MPa, while in the post-crown simulation is about 1403 MPa which is in the junction between the implant and third molar tooth.

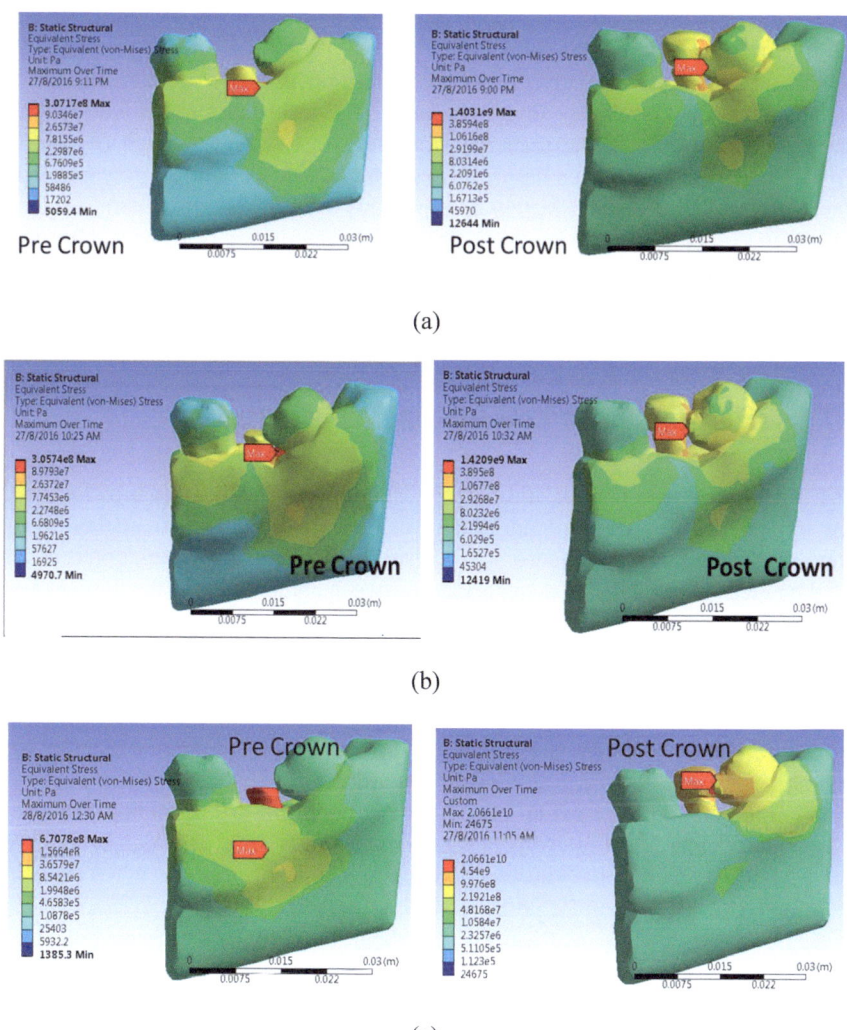

Fig. 8.2 Stress distribution on the model of low implant stability patient. **a** Vertical loading, **b** horizontal loading, and **c** removal torque at pre- and post-crown conditions

The generated stress due to horizontal loading is shown in Figure 8.2b. For the crown condition, maximum stress (0.306 GPa) is at the top of the cortical. However, in the post-crown, the maximum stress (1.42 GPa) is in the contact area between the crown and the molar tooth. The structure of the molar tooth for this patient is dipping; hence, the applied horizontal force on the crown will be affected strongly by this contact.

Quantitative assessment of stress distribution due to different loading is conducted by digitizing the stress around the implant and neighbouring teeth. The selected locations for the analysis are the side of the implant towards the second molar, the side of the implant towards the second premolar, the side of the second molar, and the side of the second premolar.

This experiment showed that the effect crown installation does not too much affect the stress distribution except in certain areas such as at the implant side towards the second premolar. In the premolar side, the stress is high enough (up to 200 MPa). This stress is exceeded limits the maximum stress for bone to grow normally. Therefore, osseointegration in this area possibly be delayed. In general, stresses around the implant that are produced from post-crown conditions are higher than pre-crown except in removal torque. In other places, the stress distribution of pre- and post-crown is almost similar.

8.1.2 Micromotion on Low Implant Stability Patient

The micromotion of implant and neighbour teeth on the model of a patient with low implant stability during loading is investigated using a similar method to the previous model. Generated micromotion resulting from loading simulation is calculated by measuring a relative displacement between components of the implant with a reference point that is in the jawbone.

Table 8.1 shows the summary of micromotion of the dental system on a model of a patient with low implant stability resulting from vertical, horizontal, and removal torque loading. Micro motion due to vertical force and horizontal force is not too much different (109.40 μm and 106.57 μm on pre-crown and 94.19 μm and 93.79 μm on post-crown). In vertical and horizontal loading, the consequence of crown installation can reduce the micro-motion of dental implants due to loading. However, if loading is in removal torque type, generated micro motion is very high in both pre- and post-crown conditions. Micromotion due to removal torque was 478.93 μm for pre-crown and 536.89 μm for post-crown conditions.

Removal of torques also affects the neighbour tooth. In the pre-crown condition, the premolar is mostly affected, while in post-crown condition, the neighbour teeth that are affected strongly are the molar tooth.

8.1 Biomechanical Assessment on Patient with Low Implant Stability

Table 8.1 Micromotion of an implant system of the patient with low implant stability

Micro motion(μm)	Simulation					
	Vertical force		Horizontal force		Removal torque	
	Pre	Post	Pre	Post	Pre	Post
Implant	109.40	94.19	106.57	93.79	478.93	536.89
Premolar	0.99	4.85	0.93	5.10	60.51	2.55
Molar	0.42	8.99	0.53	8.80	15.14	42.82

8.1.3 Summary of Result

Biomechanical assessments of patients with low implant stability have been conducted by simulating three types of loading including vertical, horizontal, and removal torque. The model that is used for this case is a bit different from the previous model. In this model, the patient has a structure of the second molar tooth that is tilted/dipping; hence, the post-crown condition where the crown is contacted with the second molar tooth affects a lot on stress transfer. The low implant stability condition has been represented by a low friction coefficient (0.2). The evaluations include the stress distribution of Von Mises stress and micro motion measurement of the dental implant and two neighbour teeth.

In the vertical loading, the trend of stress for both pre- and post-crown is like each other. The area near the implant received the stress up to 180 MPa, while the highest stress on neighbour teeth occurs on the second premolar. The second premolar tooth received stress up to 200 MPa from post-crown and 100 MPa from pre-crown. In this area, the high stress is not only located at the cortical but also can reach the trabecular. The second molar is not affected significantly by external loading either in pre- or post-crown conditions. The received stress in this area is below 6 MPa.

Horizontal loading gives a similar effect as vertical loading either in the pre- or post-crown condition. The area close to the implant received the stress of up to 200 MPa if that loading is applied to the top of the crown (post-crown) and 150 MPa if that force is applied to the top of the implant body (pre-crown). The neighbour teeth that were affected a lot by this loading are the second premolar teeth. This tooth received the stress up to 200 MPa. However, the second molar received a small stress either in pre- or post-crown condition.

In the simulation of removal torque simulation, if the torque is applied to the crown (post-crown), the generated stress is much higher than if that torque is applied to the top of the implant body directly. The area near to implant site is affected strongly, and the stress can reach up to 350 MPa. The neighbour teeth that are strongly affected by this removal torque are also the second premolar. The generated stress at this tooth can reach up to 100 MPa. The second molar tooth only received high stress in the upper part, but in the below part including the trabecula, the received stress is low.

Micromotion that was generated during the external loading is higher when vertical force and horizontal force are applied directly to the top of the implant (pre-crown condition). However, the removal torque causes a higher micro motion if the torque is located at the top of the crown (post-crown condition). The generated micro motion of dental implants due to vertical and horizontal loading is about 90–110 μm and the removal torque is up to 500 μm. The micro motions of neighbour teeth are not significant except on the applied removal torque. The micro motions of neighbour teeth due to horizontal and vertical force are below 10 μm, while the removal torque can trigger the micro motion on neighbour teeth up to 60 μm.

8.2 General Discussion on Biomechanical Evaluation of Dental Implant

It is almost impossible in current technology to measure a stress distribution from direct clinical measurement. For this reason, a Finite Element Analysis (FEA) is a powerful method that can give us the best prediction on characteristics of stress distribution on implant dental system to replace direct measurement in the clinic.

This study has been conducted to analyse the stress distribution generated by external loading in pre- and post-crown conditions. The external loading included the vertical force, horizontal force, and removal torque that might be possibly generated during daily mastication. This study started from a simple model to a complex model that involved dental implant patients with different implant stability conditions.

In the simple model, external loading either in pre- or post-crown conditions will generate high stress in the upper part of the jawbone, up to 4 mm measured from the top of the cortical, the Von Mises stress is high, and then at an area below 4 mm, the stress will decrease. The stress resulting from pre-crown conditions is higher than post-crown conditions, especially in areas below 4 mm. The stresses become higher if the loading is in the form of torque. In this case, the stress generated in the post-crown condition is higher than pre-crown. In all types of loading, the stress will be propagated from a point where loading is applied to other places. Some of the stress will be propagated through the body implant to the cortical and trabecular bone around the implant, or if there is contact between the crowns with neighbour teeth, then some of the stress will be propagated through the crown to neighbour teeth directly and then distributed to the cortical and trabecular bone around that neighbour tooth. In this experiment, the cortical bone received more stress compared with the trabecular, it can be seen from the coronal section view, that the Von Mises stress in the cortical is higher than the trabecular at the location with the same level (distance) from the top of cortical. It showed that the trabecular plays an important role in propagating the stress due to external loading.

Increasing cortical thickness means that the contact area between cortical thicknesses with an implant is increased. Different patients have different qualities of bone. When the surgery of the implant is performed, the surface characteristic around the implant site is different from patient to patient. The coupling and characteristics of bone will make the quality of contact surface between the implant and bone

8.2 General Discussion on Biomechanical Evaluation of Dental Implant

different. To evaluate the contact surface weather, and its effect on the stress distribution, an experiment with various friction coefficients was conducted. The Von Mises stresses at the same location are measured for different simulations. From four probes measurements, the experiment results showed that the friction coefficient between implant surfaces with bone correlates. Increasing in friction coefficient was followed by a decrease in Von Mises stress. Not only with the generated Von Mises stress, but the friction coefficient also correlated with the micromotion that was generated during loading. Increasing in friction coefficient between implant surfaces and bone can make the implant more stable. The micromotion generated during the simulation decreased when the friction coefficients increased. Hence, the engineering part of increasing the friction coefficient between the surfaces of an implant with the surface of the bone site can reduce the micro motion of the implant body.

Implementation analyses of FEA to estimate the stress distribution on in vivo data, three patients with different implant stability have been involved. The grades of implant stability of patients are high, moderate, and low. To represent the quality of implant stability in the simulation, the friction coefficients of patients have been adjusted differently. For a patient who has high implant stability, the friction coefficient is adjusted to 0.5, moderate implant stability is 0.4, and low implant stability is 0.2. The simulations were conducted to investigate the effect of different loading on stress distribution at pre- and post-crown conditions. The simulations are performed to simulate the possibility of generated vertical force, horizontal force, and removal torque during daily mastication. The results show that in all different patients, the stress distribution on the jawbone is concentrated highly in the cortical. In other words, the stress received by the cortical from the implant body is easier to propagate into all directions of the cortical rather than it will be transferred into the trabecular. The analysis showed that the highest stress area is in the cortical, especially around the implant site. The patterns of stress distribution on all simulations are similar to the appropriate simulation on a simple model, and only the intensity of stress is different. The generated stress on patients with low implant stability is higher than moderate and high implant stability.

The movement of a dental implant also the so-called micromotion of a dental implant is defined as a minimal displacement of an implant body relative to the surrounding tissue which is very difficult to measure directly by the currently available instrument in the clinic. In some reports, the authors showed that excessive micro-motion may interfere with the process of osseointegration of dental implants (Winter et al. 2013). They also postulated that micromotion between implant and bone must not surpass a threshold value of 150 μm for successful implant healing. In our simple model experiments, the micro motion is affected by the quality of the contact surface where the increase in friction coefficient can reduce the micro motion. In vivo, simulation with different types of loading showed that the torque can produce the highest micro motion of the implant compared with vertical or horizontal loading. Patient with low implant stability has the highest micro motion of his/her implant. In this condition, the difference between pre- and post-crown conditions is significant. It is contrary to the patient with high implant stability, the micro-motion of pre- and post-crown condition is not too much different. It is because, in the case of high implant stability, the coupling of the implant with bone is relatively higher, and the change of friction coefficient is not affected so much on the micromotion resulted.

In a biomechanical view, the success of osseointegration of dental implants can be understood from how the stress can be transferred from the body implant to the surrounding bone. Not only how the stress can be transferred, but also the intensity of the stress received in the bone surrounding the implant will affect significantly the activity of bone in remodelling/osseointegration ability. Based on the results, the highest stress can be produced by torque either in pre- or post-crown conditions. The vertical and horizontal loading can generate high stress only in the upper part of the cortical; fortunately, the generated stress by these loading is still below the critical threshold.

Reference

W. Winter, D. Klein, M. Karl, Micromotion of dental implants: basic mechanical considerations. J. Med. Eng. ID 265412 (2013)

Chapter 9
Fundamental of Three-Dimensional Bone Assessment Using CBCT from Laboratory to Clinics

9.1 Conclusion

Biomechanical assessment clinically of the dental implant system in responding to external loading such as mastication is mostly not possible. Not only because of the limitation of instruments that can be used to measure the mechanical responses but also because the experiments can influence the dental implant directly. Hence, a biomechanical assessment numerically by using Finite Element Analysis is the ultimate method to understand the behaviour of the jaw system in responding to mechanical processes such as loading during mastication. However, the workflow to achieve a reliable FEA result needs to be performed carefully.

Implementation of workflow for improvement of geometry model building based on CBCT data needs to be performed carefully. The selection of the threshold defined the precise image, especially during the separation of hard tissue from soft tissue. The geometry of each component of the jaw system can be generated using this technique. However, some difficulties might arise when the segmentation of the object needs to be performed in detail. Even though CBCT is a favourable method in dental imaging; however, in terms of image contrast, CBCT has limitations in defining the boundary of the object precisely especially if some metallic materials are placed, and the artefact appears very strong in the images. Hence, defining the boundary during the segmentation of the object based on CBCT data is still not optimum. Removing the artefact from the main data is another problem that still needs to be resolved. The proposed workflow in this research is to optimize the model generation from CBCT data through a series of segmentation steps.

Not only for generating the model of the jaw system, the CBCT data also has been used for evaluating the quality and quantity of the jaw system such as bone density. However, for some reasons, the accuracy of CBCT in determining the density is still questionable. A series of tests have been conducted to answer the unsolved problem. The test of CBCT on different angle scanning, justification of CBCT in predicting bone density by using model, and comparison with other technology (CT scan) have been conducted. Testing also has been conducted on the repeatability and accuracy of CBCT data in determining the density.

The experiment results showed that the CBCT has a dependency on angle scanning, and different angle scanning will generate different HUs of the same object. However, the change of HU is not linear with the angle changes. Compared with CT, CBCT scanning is more sensitive with different angle scanning.

As a justification of CBCT reading in estimating the density, a phantom model with density is known to be scanned. The result showed that the HU of CBCT is not linear with density. The linear regression of HU of CBCT with density has a lower coefficient correlation compared with exponential relation. As a consequence, estimating density directly from CBCT reading needs to be performed carefully, and less linearity of CBCT reading with density needs to be considered to avoid a misinterpretation on estimating the density. However, the process of density estimation from CBCT by using MIMICS software showed that there is a consistent reading in the repetition of measurement.

For monitoring of osseointegration during the treatment of dental implants, CBCT has been tested for evaluating bone density and bone quality during implant treatment. In this case, the densities are estimated by using Eq. (4.2). The density changes for every stage of monitoring were evaluated. In general, bone density decreases during the first stage of 3 months after implant placement and will increase again after 4 months from implant placement. The density decreases during the first 3 months of monitoring, it possibly related to the bone remodelling process, and increasing after 4 months is related to the osseointegration process where the bone starts to grow and Osseo is integrated into the implant.

Monitoring of the dental implant system of the patient after surgery is needed to evaluate the progress of osseointegration. Based on this study, at least three methods can be used for monitoring the progress of implant stability, direct measurement of implant stability by using RFA, monitoring the density of bone around the implant through imaging data, and numerical study by using FEA method and integrated all comprehensive information on it.

Based on samples in Hospital USM that have been used in this research, the implant stability of patients can be categorized into three classes:

1. Class 1, patients who have a significant increasing in implant stability during the healing process.
2. Class 2, patient who has the constant implant stability progress or the increasing is not significant.
3. Class 3, a patient who has negative implant stability progress, where the implant stability of the patient decreases during the healing process.

Correlation between implant stability with bone quality and quantity showed that bone density is not significantly correlated to implant stability. However, the available space for implant sites that are the bone height of the mandible and cortical thickness are correlated strongly with the implant stability of the patient. A patient who has a thicker cortical or bone height has a potential to have a high primary implant stability.

The biomechanical assessment numerically by using the FEA method showed that there is a correlation between stress intensity generated around the implant with implant stability. The stress generated during loading from patients with high and moderate implant stability is lower compared to the patients with low implant stability. Therefore, if the stresses generated around the body of the implant are high; the tendency for the bone to lose the quantity of bone is high, and hence, the optimum osseointegration is difficult to be achieved early. The potential of implant failure or low osseointegration of implant/low implant stability is higher if too much torque is subjected to the dental implant. The generated stresses are also affected by the behaviour of contact between the surface of the implant and with bone. The rougher the surface contact of implant bone, the less the generated stresses are. In other words, the higher friction coefficient between the implant and bone will reduce the generated stress around the implant body.

The mastication process can produce various loads on a new dental implant. At least three types of loading can be produced during mastication: vertical loading, horizontal loading, and torque. Depending on the stage of the implant, these loading will generate different stress distributions. In the pre-crown stage, this loading will be directed to the top of the implant body, while in the post-crown stage, this loading will be directed to the top of the crown. The response of stress distribution due to different types of loading is different. Vertical and horizontal loading will produce the stress relatively lower than torque which is safer for bone to grow. Patients with thicker cortical thickness will have some advantages in terms of the stress distribution due to loading. Because the stress will be propagated mostly in the cortical, hence the only upper part of the cortical will be affected strongly by this loading. The lower part will be safer, and bone will more easier to grow.

In terms of micromotion, the stress is related to the micromotion of dental implants. If the stress is generated strongly, the movement (micromotion) of the dental implant is also high. Generated stress and micromotion are two important factors that play an important role in supporting implant stability/osseointegration. To achieve optimum osseointegration, the area around the implant needs to be kept from overload and over-micro motion because of that loading. The condition of surface contact between the implant and with bone affects the generated micro motion of the implant during loading. The surface contact with a higher friction coefficient can reduce the generated micro motion of the implant body.

9.2 Future Work

Even HU of CBCT data can represent the density of the object, such soft or hard material can be identified easily from the image, but the quantitative prediction of density by using HU directly from CBCT is still rare. Density estimation from CBCT is an interesting task; however, the variation of HU of CBCT data somehow still affects the quality of data. In the future, these empirical equations (Eqs. 4.1–4.2) still need to be validated in many scenarios and conditions of measurement.

Some approximation of the material as homogenous material is approximated to simplify the study. It is interesting if the density of the object can be predicted from CBCT data and it can be used to perform the material assignment, and hence, the object can be assigned with variation of density (heterogeneous density distribution). In this case, the density can be predicted from the HU of CBCT data by using Eq. 4.2. This equation will bring the HU of CBCT to 3-D density model that can be used for further application such as material assignment for further FEA study.

Index

A
Abutment, 31

B
Biocompatible, 1
Biomechanical, 2–4, 7, 16, 17, 19, 63, 78, 81, 98, 99, 101
Bone density, 4, 9, 10, 13, 15, 26, 33, 44, 51, 53, 56, 57, 64, 99, 100
Bone matrix, 7
Bone quality and quantity, 8

C
Coefficient correlation, 46, 48, 53, 54, 100
Cone Beam Computed Tomography (CBCT), 2–4, 7–11, 14–16, 25–27, 32–36, 38, 39, 43–51, 53–59, 62, 65, 67–69, 75, 81, 82, 85, 91, 99–102
Cortical bone, 2, 7–9, 16, 17, 53, 68–70, 72, 81, 91, 96
Cortical thickness, 4, 13, 46, 57–59, 61–65, 75–78, 96, 100, 101
Craniofacial, 4
Crown, 4, 16, 17, 19, 26, 27, 29–33, 38, 40, 41, 55, 56, 67–75, 81–87, 88, 89, 91–98, 101
CT scan, 14, 15, 99

D
Deformation, 18, 87
Dental imaging, 7, 9, 15, 33, 35, 99

Dental implant, 1–4, 7, 10–13, 15–19, 25–28, 37, 38, 44, 51, 53, 55, 60, 63, 67, 70, 73–79, 81, 82, 84–87, 89, 91–97, 99–101
Dental impression, 29

E
Edentulous, 1, 27, 28, 30, 55, 81, 85, 91
Enamel, 14, 17, 34, 44, 46, 53

F
Field Of View (FOV), 15, 33, 35, 56
Finite Element Analysis (FEA), 2, 4, 7, 12, 13, 18, 19, 27, 37–40, 67, 70, 71, 73, 75, 78, 81, 83, 91, 92, 96, 97, 99–101
Frequency, 13, 37
Friction coefficient, 75, 76

G
Geometry, 3, 15, 19, 33, 38, 40, 62, 68, 87, 91, 99
Gingival, 10, 29, 30

H
Horizontal force, 40, 68, 71, 72, 81, 83–85, 87, 91, 94–97
Hounsfield Units (HU), 14, 15, 35, 38, 43–47, 49–54, 100–102
Hounsfield value, 10

© The Editor(s) (if applicable) and The Author(s), under exclusive license to Springer Nature Singapore Pte Ltd. 2025
M. Genisa et al., *Biomechanics of Dental Implants*, SpringerBriefs in Applied Sciences and Technology, https://doi.org/10.1007/978-981-96-6023-0

I

Implant, 1–5, 7–14, 16, 17, 19, 26–30, 32, 33, 36–38, 40, 51, 53, 55–65, 67–78, 81–87, 89, 91–98, 100, 101

Implant stability, 2–5, 7, 10–13, 26, 27, 32, 33, 36, 37, 40, 55, 56, 59–65, 74, 75, 81, 82, 84–89, 91–98, 100, 101

Insertion, 1, 9, 11–13, 28, 40, 55, 60–63, 85

Insertion torque, 9, 12, 13

Intensity, 3, 97, 98

J

Jaw, 2, 3, 43, 46, 53, 55, 58–60, 62–64, 69, 82, 83, 99

L

Loading, 1–4, 8, 10, 13, 17, 19, 27, 36, 37, 40, 41, 56, 60, 65, 67, 68, 70, 71, 73, 75, 77, 78, 83–89, 91–99, 101

M

Mandibular, 1, 4, 7, 14, 30, 31, 51, 57, 64

Material assignment, 39

Maxillary, 1, 14

Meshing, 18, 38, 39, 68, 70, 76, 86, 87, 92

Micro motion, 84, 87, 89, 94, 95, 97

Micro movement, 19, 77

MIMICS software, 33, 35–38, 44, 46, 51, 56, 57, 75, 100

Morphology, 11, 18, 30, 75

O

Osseointegration, 1–3, 8, 10, 11, 13, 26, 27, 60, 64, 65, 78, 85, 94, 97, 98, 100, 101

Osstell, 36

P

Physiological, 7

Post-crown, 3, 32, 72

Pull-out, 12

Push-out, 11, 12

R

Radiography, 14

Remodelling, 2, 7, 16, 17, 19, 62, 64

Removal torque, 40, 41, 73, 83, 88, 89, 93

Repeatability, 4, 36, 43, 44, 46, 51–53, 100

Resonance Frequency Analysis (RFA), 2, 4, 7, 11–13, 27, 29, 33, 36, 37, 60–62, 81, 91, 100

S

Screw, 28–30, 36, 78

Segmentation, 38

Shear force, 40

Simulation, 3, 12, 38, 40, 68, 70–76, 78, 83, 86, 88, 89, 91, 93–95, 97

Strain, 19, 67

Stress distribution, 2–4, 13, 40, 67–69, 71–73, 75–78, 82, 83, 87, 89, 91, 93–97, 101

Success rate, 1, 53

T

3-D image, 3, 9, 14, 75

Torsional force, 12, 40

Trabecular, 7–9, 16, 17, 19, 33, 38, 40, 46, 53, 68–70, 73, 75, 76, 78, 81, 82, 84–87, 91, 95–97

V

Von Mises stress, 19, 67, 68, 70–73, 76, 77, 82, 83, 85, 92, 95–97

X

X-ray, 14, 15, 46, 47, 50

MIX
Papier aus verantwortungsvollen Quellen
Paper from responsible sources
FSC® C105338

If you have any concerns about our products,
you can contact us on
ProductSafety@springernature.com

In case Publisher is established outside the EU,
the EU authorized representative is:
Springer Nature Customer Service Center GmbH
Europaplatz 3, 69115 Heidelberg, Germany

Printed by Libri Plureos GmbH
in Hamburg, Germany